Kaushik Parthasarathi
Editor

Molecular and Functional Insights Into the Pulmonary Vasculature

Springer

Editor
Kaushik Parthasarathi
Department of Physiology
University of Tennessee Health Science Center
Memphis, Tennessee, USA

ISSN 0301-5556 ISSN 2192-7065 (electronic)
Advances in Anatomy, Embryology and Cell Biology
ISBN 978-3-319-68482-6 ISBN 978-3-319-68483-3 (eBook)
https://doi.org/10.1007/978-3-319-68483-3

Library of Congress Control Number: 2017960938

Printed on acid-free paper

This Springer imprint is published by Springer Nature
The registered company is Springer International Publishing AG
The registered company address is: Gewerbestrasse 11, 6330 Cham, Switzerland

Advances in Anatomy, Embryology and Cell Biology publishes critical reviews and state-of-the-art surveys on all aspects of anatomy and of developmental, cellular and molecular biology, with a special emphasis on biomedical and translational topics.

The series publishes volumes in two different formats:

- Contributed volumes, each collecting 5 to 15 focused reviews written by leading experts
- Single-authored or multi-authored monographs, providing a comprehensive overview of their topic of research

228

Advances in Anatomy, Embryology and Cell Biology

More information about this series at
http://www.springer.com/series/102

Contents

1 **Transcription Factors Regulating Embryonic Development of Pulmonary Vasculature** 1
Craig Bolte, Jeffrey A. Whitsett, Tanya V. Kalin, and Vladimir V. Kalinichenko

2 **Comparative View of Lung Vascular Endothelium of Cattle, Horses, and Water Buffalo** 21
David Schneberger, Ram S. Sethi, and Baljit Singh

3 **The Pulmonary Vascular Barrier: Insights into Structure, Function, and Regulatory Mechanisms** 41
Kaushik Parthasarathi

4 **Pulmonary Endothelial Cell Apoptosis in Emphysema and Acute Lung Injury** 63
Eboni Chambers, Sharon Rounds, and Qing Lu

Chapter 1
Transcription Factors Regulating Embryonic Development of Pulmonary Vasculature

Craig Bolte, Jeffrey A. Whitsett, Tanya V. Kalin, and Vladimir V. Kalinichenko

Abstract Lung morphogenesis is a highly orchestrated process beginning with the appearance of lung buds on approximately embryonic day 9.5 in the mouse. Endodermally derived epithelial cells of the primitive lung buds undergo branching morphogenesis to generate the tree-like network of epithelial-lined tubules. The pulmonary vasculature develops in close proximity to epithelial progenitor cells in a process that is regulated by interactions between the developing epithelium and underlying mesenchyme. Studies in transgenic and knockout mouse models demonstrate that normal lung morphogenesis requires coordinated interactions between cells lining the tubules, which end in peripheral saccules, juxtaposed to an extensive network of capillaries. Multiple growth factors, microRNAs, transcription factors,

C. Bolte
Center for Lung Regenerative Medicine, Perinatal Institute, Cincinnati Children's Research Foundation, Cincinnati, OH, USA

Division of Pulmonary Biology, Cincinnati Children's Research Foundation, Cincinnati, OH, USA

J.A. Whitsett
Division of Pulmonary Biology, Cincinnati Children's Research Foundation, Cincinnati, OH, USA

Division of Developmental Biology, Perinatal Institute, Cincinnati Children's Research Foundation, Cincinnati, OH, USA

T.V. Kalin
Division of Pulmonary Biology, Cincinnati Children's Research Foundation, Cincinnati, OH, USA

V.V. Kalinichenko (✉)
Center for Lung Regenerative Medicine, Perinatal Institute, Cincinnati Children's Research Foundation, Cincinnati, OH, USA

Division of Pulmonary Biology, Cincinnati Children's Research Foundation, Cincinnati, OH, USA

Division of Developmental Biology, Perinatal Institute, Cincinnati Children's Research Foundation, Cincinnati, OH, USA
e-mail: Vladimir.Kalinichenko@cchmc.org

K. Parthasarathi (ed.), *Molecular and Functional Insights Into the Pulmonary Vasculature*, Advances in Anatomy, Embryology and Cell Biology 228,
https://doi.org/10.1007/978-3-319-68483-3_1

and their associated signaling cascades regulate cellular proliferation, migration, survival, and differentiation during formation of the peripheral lung. Dysregulation of signaling events caused by gene mutations, teratogens, or premature birth causes severe congenital and acquired lung diseases in which normal alveolar architecture and the pulmonary capillary network are disrupted. Herein, we review scientific progress regarding signaling and transcriptional mechanisms regulating the development of pulmonary vasculature during lung morphogenesis.

1.1 Introduction

The mature lung is a complex organ consisting of a multiplicity of cell types organized in a precisely controlled pattern during morphogenesis. Perturbations of pulmonary developmental programs cause congenital malformations that are a common cause of morbidity and mortality in infancy. The pulmonary vasculature, which provides efficient gas exchange between air and blood, is composed of the bronchial and pulmonary vascular networks (Bridges and Weaver 2006). The bronchial circulation is directly supplied by the systemic circulation and is derived from the aorta. The bronchial circulation is formed in close proximity to the conducting airways and provides oxygen and nutrient-rich blood to pulmonary tissues. The pulmonary vasculature is distinct from the systemic circulation, delivering deoxygenated blood to the lung from the right ventricle of the heart via the pulmonary artery and returning oxygen-rich blood to the left atrium via the pulmonary vein. The pulmonary vascular network can be subdivided into proximal and peripheral circulations. The proximal circulation consists of pulmonary veins and arteries. The peripheral pulmonary circulation consists of a mesh of microvessels and capillaries which create the vascular bed within the alveolar region of the peripheral lung where gas exchange occurs. A complex lymphatic system is also present in the lung that serves to maintain tissue fluid clearance. The development of the pulmonary capillary network will be the primary focus of this chapter.

Morphogenesis of the embryonic pulmonary vasculature occurs via two main processes: vasculogenesis (de novo formation of blood vessels) and angiogenesis (branching of preexisting blood vessels). Vasculogenesis is highly active during early stages of embryogenesis, as endothelial progenitor cells (angioblasts) differentiate into endothelial cells. Angiogenesis is the major process by which the pulmonary vasculature develops. Both angiogenesis and vasculogenesis require precise transcriptional regulation and signaling mediated by Vascular Endothelial Growth Factor (VEGF). Formation of embryonic vasculature depends on signaling through ANGIOPOIETIN/TIE2, PDGF, PI3K/AKT, TGF-β, SHH, WNT, and NOTCH, as well as transcription factors FOXF1, FOXM1, FOXC1/2, ETV2, HAND1, MEF2C, PROX1, HEY1/2, COUP-TFII, TBX4, SNAIL, GATA, SOX, and KLF (reviewed in De Val and Black 2009; Arora and Papaioannou 2012; Tiozzo et al. 2012). Temporal and spatial coordination of these signaling and transcriptional programs generates the extensive pulmonary vascular network required for lung function.

1.2 Lung Morphogenesis

1.2.1 Structural Development of the Respiratory Tree and Pulmonary Vasculature

The pulmonary vascular tree develops in close proximity to the epithelial lining of the lungs in a process mediated by crosstalk between respiratory epithelial cells and those of the pulmonary mesenchyme. Lung development in mice proceeds through distinct anatomic stages defined as: embryonic [prior to embryonic day 9.5 (E9.5)], pseudoglandular (E9.5–16.5), canalicular (E16.6–17.4), saccular (E17.5–postnatal day 5 (P5), and alveolar (P5–P20) (Fig. 1.1) (Costa et al. 2001; Cardoso and Lu 2006; Morrisey and Hogan 2010; Warburton et al. 2010; Herriges and Morrisey 2014; Kool et al. 2014). Epithelial tubules in the embryonic lung undergo branching morphogenesis during the pseudoglandular stage to generate a primitive tree-like network of tubules (Metzger et al. 2008; Morrisey and Hogan 2010; Warburton et al. 2010). Three modes of branching sequentially transform the lung buds into a more complex organ. Domain branching creates a series of new buds which arise from the bronchial stalk at specified distances, generating a series of smaller bronchi. Subsequently, the tip of the growing buds divides by either planar or orthogonal bifurcation based on a series of clues to prevent steric hindrance of the growing respiratory tree (Metzger et al. 2008; Morrisey and Hogan 2010; Warburton et al. 2010). During the canalicular stage, branching morphogenesis continues in the periphery of the lung thereby increasing the number of subdivisions of the tubular network. Terminal bronchioles give rise to respiratory bronchioles and alveolar ducts as airway epithelium differentiates (Warburton et al. 2010). Blood vessels grow with branching of the airways, increasing the complexity of the pulmonary vascular network (Morrisey and Hogan 2010; Warburton et al. 2010). The saccular stage involves thinning of the interstitium and differentiation of alveolar type I and type II epithelial cells in association with rapid capillary growth and increasingly close apposition of endothelial-lined capillaries and epithelial-lined alveolar saccules. The complexity of the pulmonary lymphatic network increases with advancing development (Warburton et al. 2010). While gas exchange can occur at the saccular stage, further septation of the saccules to form alveoli increases surface area and efficiency of gas exchange. Formation of secondary septae creates alveoli lined by type I and type II epithelial cells in close apposition to the capillary network allowing efficient diffusion of gases between the pulmonary circulation and the alveolar space (Morrisey and Hogan 2010; Warburton et al. 2010). Although there is considerable knowledge regarding the

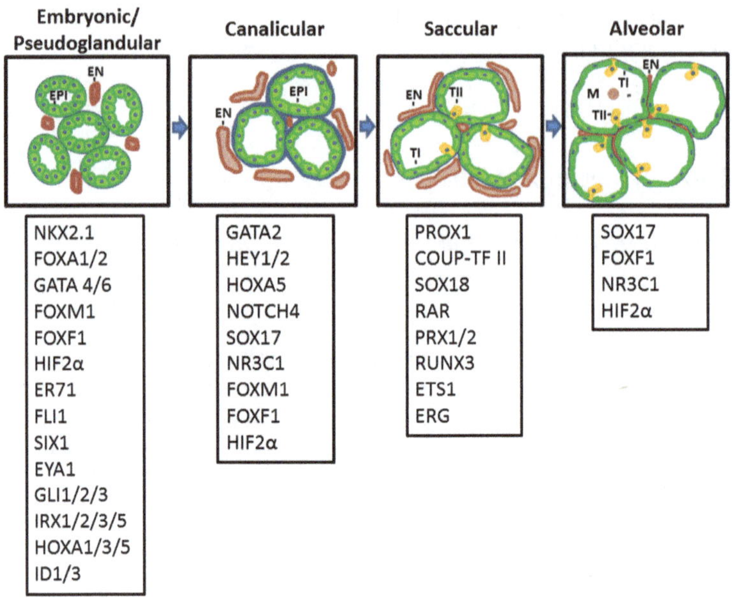

Fig. 1.1 Stages and transcriptional regulators of lung vascular development. The murine lung progresses through four developmental stages in utero and final fine-tuning occurs postnatally during the alveolar stage. Images represent the appearance of the epithelial tree as well as developing pulmonary vasculature during the pseudoglandular, canalicular, saccular, and alveolar stages of lung development. The major transcription factors regulating development of the pulmonary vasculature are listed for each stage. Abbreviations: *EN* endothelium, *EPI* epithelium, *TI* type I pneumocyte, *TII* type II pneumocyte, *M* macrophage

signaling pathways regulating development and differentiation of respiratory epithelium, development of pulmonary vasculature is not well understood.

1.2.2 Signaling Pathways Regulating Lung Morphogenesis

Multiple growth factors, transcription factors, and their associated signaling cascades are involved in formation of the pulmonary vasculature (Table 1.1). WNT and BMP signaling pathways are required for specification of the lung (Herriges and Morrisey 2014). Mice lacking WNT2/2b or β-CATENIN fail to generate a lung field, whereas over-activation of this pathway expands the specified lung field into the anterior foregut (Goss et al. 2009). Loss of BMP4 or BMP receptors 1A/B results in embryonic lethality as endoderm fails to form a trachea or, subsequently, the bronchial tree (Domyan et al. 2011). Formation of the gut tube also requires retinoic acid (RA) signaling through its receptors (RARs) (Cardoso and Lu 2006; Warburton et al. 2010). RA antagonism in mice from conception results in mid-gestation lethality and loss of lung field specification (Niederreither et al. 1999; Desai et al.

Table 1.1 Signaling molecules and transcription factors regulating pulmonary vascular development as determined by genetic mouse models

Gene	Mouse model	Vascular phenotype	Reference
Signaling molecules			
Vegf	*SPC-Vegf*	Abnormal pulmonary vascular development	Zeng et al. (1998)
Vegf	*Lyve-1-Cre Vegfr2*$^{-/-}$	Lymphatic hypoplasia, decreased blood vessel development in yolk sac, liver, and lung	Dellinger et al. (2013)
Tgf-β	*SPC-Tgf-β1 TG* (constitutively active)	Developmental arrest in pseudoglandular stage, increased vascular density	Zeng et al. (2001)
Shh	*Tie2-Cre Shh*$^{-/-}$	Simplified pulmonary vasculature	van Tuyl et al. (2007)
Wnt	*Wnt2*$^{-/-}$*;Wnt2b*$^{-/-}$	Lung agenesis, lack of lung specification	Goss et al. (2009)
Notch	*Tie2-tta Tre-Notch4*	Arterial-venous shunts	Miniati et al. (2010)
Bmp	*Shh-Cre Bmpr1a*$^{fl/+}$*; Bmpr1b*$^{-/-}$	Die at birth, tracheal agenesis and aberrant bronchi	Domyan et al. (2011)
Pten	*Dermo-Cre Pten*$^{-/-}$	Reduced pulmonary capillary network, ACD-like syndrome	Tiozzo et al. (2012)
Transcription factors			
Foxf1	*Foxf1*$^{+/-}$	Hypoplastic lung, fusion of right lung lobes, tracheal narrowing and fistula, ACD-like syndrome	Mahlapuu et al. (2001)
Foxm1	*Foxm1*$^{-/-}$	Late embryonic lethal, hypertrophy of αSMA^{+} cells, defective peripheral pulmonary capillary formation	Kim et al. (2005)
Foxm1	*Smmhc-Cre Foxm1*$^{-/-}$	Perinatal lethal, pulmonary hemorrhage	Ustiyan et al. (2009)
Foxc2	*Foxc2*$^{-/-}$	Abnormal lymphatic patterning, agenesis of valves	Petrova et al. (2004)
Foxo1	*Tie2-Cre Foxo1*$^{-/-}$	Endothelial cell hyperplasia, vessel thickening	Wilhelm et al. (2016)
Prox1	*Prox1*$^{-/-}$	Lethal by E15, complete lack of lymphatic capillary network	Wigle and Oliver (1999)
Hey1/2	*Hey1*$^{-/-}$*; Hey2*$^{-/-}$	Lethal after E9.5, impaired vascular remodeling and hemorrhage	Fischer et al. (2004)
Coup-tf II	*Coup-tf II*$^{-/-}$	Die circa E10, defective remodeling of primitive capillary plexus	Pereira et al. (1999)
Sox17	*Sox17*$^{-/-}$	Aberrant heart looping, enlarged cardinal vein, anterior dorsal aorta defects	Sakamoto et al. (2007)
Sox17	*Dermo1-Cre Sox17*$^{-/-}$	Die <3 weeks, decreased pulmonary microvasculature, alveolar simplification	Lange et al. (2014)

(continued)

Table 1.1 (continued)

Gene	Mouse model	Vascular phenotype	Reference
Sox18	*Sox18*$^{-/-}$	Lethal around E14.5, lack of lymphatic endothelial cell differentiation	Francois et al. (2008)
Fli1	*Fli1*$^{-/-}$	Blood vessel leak/hemorrhage	Spyropoulos et al. (2000)
Eya1	*Eya1*$^{-/-}$	Blood congestion, decreased vascular smooth muscle, blood vessel herniation	El-Hashash et al. (2011a)
Six1	*Six1*$^{-/-}$	Blood congestion, decreased vascular smooth muscle, blood vessel herniation	El-Hashash et al. (2011b)
Id1/3	*Id1*$^{-/-}$; *Id3*$^{-/-}$	Lack of branching/sprouting of blood vessels into neuroectoderm	Lyden et al. (1999)
Er71	*Er71*$^{-/-}$	Loss of Flk-1/endothelial cells	Lee et al. (2008)
Erg	*Tie2-Cre Erg*$^{-/-}$	Immature/disorganized vascular plexus	Birdsey et al. (2015)
Hoxa5	*Hoxa5*$^{-/-}$	Decreased branching, alveolar thickening, no surfactant	Aubin et al. (1997)
Nr3c1	*Col1-Cre GR*$^{-/-}$	Die at birth with respiratory distress and cyanosis, lung at pseudoglandular stage	Habermehl et al. (2011)
Nr3c1	*Dermo1-Cre GR*$^{-/-}$	Die after birth from cyanosis, lung atelectasis, developmental arrest in canalicular stage	Li et al. (2013)
Prx1	*Prx1*$^{-/-}$	Die postnatal due to cyanosis, hypoplastic lung, reduced vWF+ blood vessels	Ihida-Stansbury et al. (2004)
Runx3	*Runx3*$^{-/-}$	Die within 24 h of birth; poor alveolar expansion/vascularization	Lee et al. (2014a)

2006). RARβ signaling is essential for lung field specification while RARα and RARγ are also required for branching lung morphogenesis (McGowan et al. 2000; Desai et al. 2006). The FGF family is also critical for lung development as multiple FGFs including FGF10, FGF9, FGF18, FGF8, and receptors FGFR2B and FGFR3/4 regulate formation of the lung buds (Cardoso and Lu 2006), branching of the bronchioles (Morrisey and Hogan 2010; Herriges and Morrisey 2014), and influence development of the pulmonary vasculature (White et al. 2006). Hedgehog signaling is critical for development of the lung and other organs. SHH is derived from the endoderm and mediates expression of *Gli* genes in the mesenchyme as well as *Fgf* in the mesoderm (Warburton et al. 2000; Kumar et al. 2005). SHH is essential for lung bud formation, branching morphogenesis, and development of pulmonary vasculature (van Tuyl et al. 2007; Morrisey and Hogan 2010). *Shh*$^{-/-}$ mice have significant branching defects, and conditional deletion of *Shh* from endothelial and hematopoietic lineages (*Tie2-Cre Shh*$^{-/-}$) decreases complexity of the pulmonary vascular bed (van Tuyl et al. 2007). Conversely, TGF-β negatively regulates lung development as a constitutively active mutant form of TGF-β_1 leads to developmental arrest in the pseudoglandular stage with poor sacculation and epithelial differentiation (Zeng et al. 2001). The PDGF family, VEGF, and retinoic acid influence alveolarization including formation of the capillary network in the alveolar region (Kumar et al. 2005). VEGF, in particular, is critical for development of lung vasculature and, indeed, for vascular formation in general. Targeted disruption of the *Vegf* gene in mice produces an embryonic lethal phenotype due to impaired blood-island formation and delayed differentiation of endothelial cells, causing abnormal blood vessel development (Carmeliet et al. 1996; Ferrara 1996). VEGF signals through tyrosine kinase receptors FLK1 and FLT1 that are expressed in endothelial cells and their progenitors. *Flk1*$^{-/-}$ mice exhibit an embryonic lethal phenotype due to disruption of vasculogenesis and loss of angioblast cells from blood islands (Shalaby et al. 1995), while *Flt1*$^{-/-}$ embryos fail to form mature blood vessels (Fong et al. 1995). Lung-specific overexpression of VEGF (*SPC-Vegf* transgenic mice) causes neonatal lethality associated with increased pulmonary vascular development and disrupted epithelial differentiation (Zeng et al. 1998). VEGF has also been shown to influence expression of *Etv2*, a transcription factor essential for development of blood vessels in the lungs and other organs (Casie Chetty et al. 2017). NR3C1, glucocorticoid receptor, signaling is critical for lung maturation as *Nr3c1*$^{-/-}$ mice die after birth due to respiratory failure and corticosteroid administration to late term or preterm babies is a long standing medical practice (Habermehl et al. 2011). Recently, microRNA (miR) have been shown to regulate lung vascular development, as silencing of either DICER or DROSHA reduces endothelial sprouting and epithelial-specific loss of *Dicer* decreases epithelial branching (Marcelo et al. 2013; Herriges and Morrisey 2014). miR-29 is the most common microRNA in adult lungs, with highest expression in the distal vasculature (Cushing et al. 2015). Disruption of miR-29 inhibits vascular smooth muscle development with increased expression of *Klf4*, *Pdgfrb*, *Fbxo32*, and *Foxo3a* (Cushing et al. 2015). The microRNA cluster miR-17-92 is highly expressed in early lung development and its overexpression driven by the *Spc* promoter results in abnormal lung development and lethality (Lu et al. 2007). Disruption of the miR-17-92 cluster causes perinatal lethality due to hypoplastic lung (Ventura et al. 2008). miR clusters miR302-367 and miR449/34 as well as miR375 have been shown to regulate cell differentiation in the later stages of lung development (Herriges and Morrisey 2014). Among microRNAs, miR-126 and miR-221 are pro-angiogenic and miR-27b promotes endothelial cell tip specification and branching of the vascular plexus (Marcelo et al. 2013). Long noncoding RNAs (lncRNAs) are another emerging player in lung development. Mice deficient in the lncRNA *Fendrr* die at birth with hypoplastic lungs and enlarged alveolar saccules (Sauvageau et al. 2013).

1.3 Transcriptional Regulation of Lung Development

1.3.1 Embryonic Stage

In order to control a process as complex as lung organogenesis, precise temporal and spatial regulation of gene expression is required. Homeobox transcription factor NKX2.1 is the earliest known marker of the primitive lung field, marking respiratory epithelial cells in the primordial lung buds. *Nkx2.1*-null mice die from respiratory failure at birth caused by severe lung hypoplasia (Kimura et al. 1996). *Nkx2.1* is regulated by FOXA2, a member of the Forkhead box (Fox) family of transcription factors, and by GATA6 (Ikeda et al. 1996; Shaw-White et al. 1999; Costa et al. 2001; Kumar et al. 2005). *Nkx2.1* expression depends on the canonical WNT/β-CATENIN pathway and BMP signaling. Disruption of either FOXA2 or GATA6 pathways blocks expression of *Nkx2.1* and inhibits lung specification (Herriges and Morrisey 2014). GATA4 also plays a role in the early specification of the lung field (Kumar et al. 2005). Mice lacking *Foxa1/2*, *Gata4*, or *Gata6* die at early stages of embryogenesis due to multiple developmental abnormalities, including heart defects, loss of extra-embryonic endoderm, and severely underdeveloped gut tube (Costa et al. 2001; Cardoso and Lu 2006). All these transcription factors are expressed in foregut endoderm and endoderm-derived lung epithelial cells, suggesting they control epithelial cell fate decisions during lung specification.

1.3.2 Pseudoglandular Stage

Branching morphogenesis in the pseudoglandular stage of lung development is associated with increased angiogenesis. The ETS transcription factors ER71 and FLI1 play important roles early in pulmonary vascular development (Fig. 1.1). *Fli1* is expressed in endothelial cells of the developing vasculature, and deletion of *Fli1* in mouse causes death before E12.5 related to vascular leak (Lelievre et al. 2001). In the mouse, *Er71* can be detected as early as E7, and its expression is regulated by VEGF, BMP, NOTCH, and WNT signaling pathways (Kataoka et al. 2011). *Er71*$^{-/-}$ mice die at E9.5 due to lack of endothelial cells, decreased *Fli1* and *Gata6* expression, and loss of *Flk1* (Lee et al. 2008). Both ER71 and GATA2 are critical for differentiation of hemangiogenic mesoderm into endothelial cells as opposed to hematopoetic cell lineages (Kataoka et al. 2011; Park et al. 2013). The homeobox transcription factor SIX1 and its transcriptional co-activator EYA1 are important regulators of lung vascular development. Deletion of either *Eya1* or *Six1* causes lung hypoplasia associated with decreased branching (Table 1.1), as well as disruption of FGF10 and SHH signaling (El-Hashash et al. 2011a, b). Both FGF10 and SHH are critical for epithelial–mesenchymal interactions during formation of the lung (van Tuyl et al. 2007; Morrisey and Hogan 2010). Deletion of *Eya1* or *Six1* causes vascular defects including decreased vascular smooth muscle and herniation of the blood vessels

(El-Hashash et al. 2011b). SHH plays critical roles during lung development, mediated by the GLI family of zinc finger transcription factors. Deletion or mutation of either *Gli2* or *Gli3* results in failure of normal lobulation, generating a simplified lung (Grindley et al. 1997; Motoyama et al. 1998; Costa et al. 2001). While mice with mutant *Gli1* or compound *Gli1/3* mutations have no lung phenotype, compound mutations in *Gli1/2* or *Gli2/3* cause severe lung hypoplasia (Grindley et al. 1997; Motoyama et al. 1998; Park et al. 2000; Costa et al. 2001). Although no vascular phenotype has been described in *Gli*-deficient mice, GLI2 and GLI3 inhibit FGF10, a growth factor essential for pulmonary vascular development (Warburton et al. 2000). Multiple isoforms of the IRX homeobox transcription factors, including *Irx1*, *Irx2*, *Irx3*, and *Irx5*, are expressed in the lung and play redundant roles during lung formation and branching morphogenesis. Simultaneous deletion of these *Irx*'s inhibits epithelial branching in vitro and inhibits differentiation of airway epithelial cells (van Tuyl et al. 2006). Decreased α-smooth muscle actin surrounding pulmonary vessels was observed after inhibition of the *Irx* transcription factors, supporting a role for IRX in the development of vascular smooth muscle (van Tuyl et al. 2006). Deletion of *Hoxa5* inhibited branching morphogenesis, caused alveolar thickening, and inhibited surfactant production (Aubin et al. 1997; Costa et al. 2001; Jones 2003); whereas deletion of either *Hoxa1* or *Hoxa3* caused lung hypoplasia and respiratory failure at birth (Chisaka and Capecchi 1991; Lufkin et al. 1991; Jones 2003). Consistent with a critical role of BMP signaling during pulmonary vascular development, mice lacking the BMP/SMAD targets ID (inhibitor of differentiation and DNA binding 1 and 3) died before E13.5 and exhibited vascular defects and decreased branching morphogenesis (Lyden et al. 1999; Park et al. 2013).

1.3.3 Canalicular Stage

During the canalicular stage of lung development, branching morphogenesis continues in the terminal saccules, which will later undergo sacculation to form alveoli. Vascular development is highly active during this stage, as an extensive capillary network is formed. NOTCH signaling has emerged as a major regulator of pulmonary vascular development. NOTCH signaling stabilizes angiogenic sprouts which are essential for expansion of the vascular network (Lee et al. 2014b). NOTCH signaling also influences expression of *Vegfa*, which is critical for vascular development (Crivellato 2011). Loss of VEGFA activity has been shown to result in decreased numbers of pulmonary capillaries and reduced alveolarization (Yun et al. 2016). Endothelial-specific expression of constitutively active *Notch4* caused arterial-venous shunts (Miniati et al. 2010). In support of a critical role of NOTCH in vascular development, embryos deficient in NOTCH target genes *Hey1/2* die during the early stage of lung development with impaired vascular remodeling (Fischer et al. 2004; Park et al. 2013) (Table 1.1). Furthermore, NOTCH inhibits *Sox17*, a transcription factor critical for vascular development (Lee et al. 2014b). *Sox17*-null mice die at E10.5 (Kanai-Azuma et al. 2002), while endothelial-specific deletion of *Sox17* disrupted formation of the pulmonary microvasculature and caused alveolar simplification (Lee et al. 2014b). Conversely, endothelial overexpression of *Sox17* enhances angiogenic sprouting (Lee et al. 2014b). Mice with mesenchymal-specific *Sox17* deletion mediated by *Dermo1-Cre* die within 3 weeks of birth with a significant decrease in pulmonary microvasculature and alveolar simplification (Lange et al. 2014). Forkhead box (Fox) transcription factors are also important regulators of pulmonary vascular formation. Severe defects in pulmonary mesenchyme and undeveloped capillaries were observed in *Foxm1*$^{-/-}$ embryos causing death in utero (Wang et al. 2003; Kim et al. 2005), demonstrating FOXM1 is critical for lung development. Perturbations in *Foxm1* expression have profound effects on multiple organ systems, including the cardiovascular, gastrointestinal, and hematopoietic systems (Wang et al. 2003; Ren et al. 2010; Bolte et al. 2011, 2012; Sengupta et al. 2013). *Foxm1* deficiency inhibits development of Clara and goblet cells in airway epithelium and disrupts normal epithelial patterning in the developing lung (Wang et al. 2010, 2012; Ustiyan et al. 2012, 2016). FOXM1 stimulates proliferation of endothelial cells during formation of pulmonary vasculature and accelerates cell cycle progression in normal and neoplastic cells during tissue repair, carcinogenesis, and inflammation (Liu et al. 2011; Balli et al. 2012; Ren et al. 2013; Cheng et al. 2014; Wang et al. 2014; Gao et al. 2015; Kalinichenko and Kalin 2015; Xia et al. 2015). Conditional deletion of *Foxm1* from smooth muscle cells using *smooth muscle myosin heavy Cre* (*Smmhc-Cre Foxm1*$^{-/-}$) resulted in decreased vascular muscle formation and pulmonary hemorrhage (Ustiyan et al. 2009), supporting an important role for FOXM1 in pulmonary vascular development. FOXO1 has been shown to inhibit endothelial proliferation as mice deficient for endothelial *Foxo1* (*Tie2-Cre Foxo1*$^{-/-}$ mice) have increased rates of proliferation resulting in thicker vessels than control mice (Wilhelm et al. 2016). *Foxf1* and *Foxf2* are expressed in mesenchyme-derived cells of the lung and are regulated by SHH (Aitola et al. 2000; Bolte et al. 2015; Xu et al. 2016; Milewski et al. 2017). While *Foxf1*$^{-/-}$ mice die before E9.5 related to defects in the yolk sac and allantois (Mahlapuu et al. 2001; Kalinichenko et al. 2003b), *Foxf1*$^{+/-}$ mice frequently die at birth due to disruption of pulmonary vascular formation, a syndrome similar to alveolar capillary dysplasia (ACD) (Kalinichenko et al. 2001; Sen et al. 2014). *Foxf1*$^{+/-}$ mice are also more susceptible to lung injury indicating the importance of FOXF1 in lung injury and repair (Kalin et al. 2008; Cai et al. 2016; Pradhan et al. 2016). Mesenchymal-specific inactivation of *Pten* resulted in death at birth, decreased *Foxf1* expression, and caused pathology consistent with findings in ACD, thus implicating the PI3K/AKT pathway in development of the pulmonary vasculature (Tiozzo et al. 2012). Interestingly, overexpression of *Foxf1* in transgenic mice also causes perinatal lethality with vascular hemorrhage, immature lungs, and disrupted capillary network (Dharmadhikari et al. 2016), indicating pulmonary vascular development is dependent on dosage of the *Foxf1* gene. HIF1α and HIF2α regulate the expression of *Vegf* and thus influence normal formation of the pulmonary vasculature (Jones 2003; Galambos and deMello 2007; Park et al. 2013; Kool et al. 2014; Tibboel et al. 2015). Mice with mesenchyme-specific deletion of the *Nr3c1* nuclear receptor failed to progress from the canalicular to the saccular stage of lung development (Li et al. 2013). These mice were cyanotic at birth and died shortly afterwards with lung morphology consistent with the canalicular stage of development (Li et al. 2013).

1.3.4 Saccular/Alveolar Stages

Considerable remodeling of lung structure occurs during the saccular and alveolar stages of lung development as the number of terminal saccules increases, complexity of the vascular network expands, and development of the pulmonary lymphatic system occurs (Costa et al. 2001; Cardoso and Lu 2006; Morrisey and Hogan 2010; Warburton et al. 2010; Herriges and Morrisey 2014; Kool et al. 2014). This period is defined by increased specialization and differentiation of respiratory epithelial cells, which differ across the proximal–peripheral gradient of the airways. PROX1 and COUP-TF II mediate formation of lymphatic vasculature and are actively expressed during the saccular stage of lung development (Pereira et al. 1999; Wigle and Oliver 1999; Galambos and deMello 2007). SOX18, known to be a critical regulator of endothelial venous cell differentiation, stimulates formation of lymphatic vessels (Francois et al. 2008; Park et al. 2013). Deletion of *Foxc2* inhibited lymphatic vessel formation (Petrova et al. 2004; Park et al. 2013), and mice lacking both *Foxc1* and *Foxc2* die during early mouse embryogenesis with defective vascular remodeling (Kume et al. 2001; Park et al. 2013). COUP-TF II blocks retinoic acid receptor signaling (Kimura et al. 2002), suggesting RAR signaling directs endothelial cells towards a vascular versus lymphatic cell fate. VEGF signaling also influences lymphatic development as deletion of *Flk1* from lymphatic vasculature (*Lyve-1-Cre Vegfr2*$^{-/-}$ mice) was sufficient to cause embryonic death at E14.5. *Lyve-1-Cre Vegfr2*$^{-/-}$ mice presented with lymphatic hypoplasia as well as decreased blood vessel formation in the yolk sac, liver, and lungs (Dellinger et al. 2013). Paired-class homeobox transcription factors PRX1 and PRX2 have been linked to pulmonary hypertension due to a role in smooth muscle proliferation (Jones 2003), and *Prx1*$^{-/-}$ mice have been shown to have decreased lung vascular smooth muscle development (Ihida-Stansbury et al. 2015). *Prx1*$^{-/-}$ mice die soon after birth from cyanosis and exhibit reduced blood vessel formation and decreased expression of *Flk-1* and *Vcam-1*, molecules essential for blood vessel formation and integrity (Jones 2003; Ihida-Stansbury et al. 2004). *Runt-related transcription factor 3* knockout mice (*Runx3*$^{-/-}$) die within the first day of life and demonstrate vascular abnormalities (Lee et al. 2014a). RUNX3 inhibits *Pecam-1*, *Vegf*, and *von Willebrand factor* expression (Lee et al. 2014a), suggesting a role in endothelial cell growth and/or maintenance. ETS1, expressed by endothelial cells during angiogenesis, regulates the expression of numerous genes involved in vascular development including *Tie1/2*, *Flt-1*, *Flk-1*, and *VE-cadherin*. Remarkably, *Ets1*$^{-/-}$ mice have no abnormalities in the vascular system, likely related to redundancy of expression of other ETS transcription factors including FLI1 and

ERG1, both known to influence expression of *VE-cadherin* and *Flt1* (Lelievre et al. 2001; Grzenda et al. 2013). ERG in particular has been shown to be critical for vascular development as endothelial deletion of *Erg* (*Tie2-Cre Erg*$^{-/-}$) causes death before E9.5, prior to lung vascular development, and mice heterozygous for endothelial *Erg* have decreased levels of *VE-cadherin* in lung endothelial cells (Birdsey et al. 2015).

1.4 Foxf1 Transcription Factor Is a Critical Regulator of Pulmonary Vascular Development

Alveolar Capillary Dysplasia with Misalignment of Pulmonary Veins (ACD/MPV) is one of the most severe human congenital disorders. This syndrome, which occurs in neonates and in early infancy, is associated with impaired lobulation, loss and displacement of pulmonary capillaries, abnormalities in formation of pulmonary arteries and malposition of veins, which leads to pulmonary hypertension, lung edema, cyanosis, and respiratory failure (Bishop et al. 2011). ACD/MPV patients frequently present with other congenital malformations of the genitourinary, gastrointestinal, and cardiovascular systems. The severity of congenital abnormalities and progressive respiratory insufficiency in ACD/MPV infants usually results in death shortly after birth (Bishop et al. 2011). Recently, heterozygous deletions and point mutations in the *FOXF1* gene locus were identified in up to 50% of ACD/MPV patients (Stankiewicz et al. 2009; Dharmadhikari et al. 2015). These clinical studies demonstrate a critical role for FOXF1 in development of the capillary network of the human lung.

FOXF1 (HFH-8 or FREAC-1) is a member of the Forkhead Box (Fox) family of transcription factors which share an evolutionary conserved Winged helix/Forkhead DNA binding domain. *Foxf1* is expressed in endothelial and mesenchymal cells of yolk sac, placenta, and allantois, and *Foxf1*$^{-/-}$ mice die in the early embryonic period at approximately E8.5 with vascular abnormalities in the extra-embryonic mesoderm (Peterson et al. 1997; Mahlapuu et al. 2001; Kalinichenko et al. 2002). *Foxf1* haploinsufficiency in mice causes an alveolar capillary dysplasia-like syndrome, associated with lung hypoplasia, fusion of lung lobes, loss of pulmonary capillaries, and various developmental defects in the mesenchyme of gallbladder, esophagus, and trachea (Kalinichenko et al. 2001, 2002; Mahlapuu et al. 2001). A subset of *Foxf1*$^{+/-}$ mice die at birth, in association with ACD and progressive respiratory insufficiency (Kalinichenko et al. 2001). The remarkable similarities between mouse and human phenotypes associated with mutations in *Foxf1* indicate these mice provide a useful model to study the pathogenesis of ACD and to develop therapeutics to intervene with this congenital pediatric disorder. In fact, *Foxf1* promoter and coding regions have been found to be highly conserved from mouse to humans (Kim et al. 2005); however, *Foxf1* haploinsufficiency in mice is less severe than the syndromes seen in humans, since 50–70% of *Foxf1*$^{+/-}$ mice survive after birth (Kalinichenko et al. 2001). Although these mice usually reach adulthood, they are susceptible to lung injury, pulmonary hemorrhage, and exhibit abnormal lung and liver repair (Kalinichenko et al. 2002, 2003a).

Foxf1 is a downstream target of SHH signaling and *Shh*$^{-/-}$ mice lack *Foxf1* expression (Mahlapuu et al. 2001). SHH-mediated activation of *Foxf1* occurs through direct binding of GLI transcription factors to the *Foxf1* gene promoter region (Madison et al. 2009; Hoffmann et al. 2014). Loss of GLI binding sites in the *FOXF1* gene locus were found in ACD/MPV patients (Szafranski et al. 2013), emphasizing the importance of the SHH/GLI/FOXF1 signaling pathway in the regulation of pulmonary vascular development. FOXF1 plays a critical role in migration of endothelial and mesenchymal cells through activation of *Integrin β3* and *Notch-2* (Kalinichenko et al. 2004; Malin et al. 2007). Mice with *Foxf1*-floxed alleles were recently generated (Hoggatt et al. 2013; Ren et al. 2014). Inactivation of *Foxf1* in endothelial cells using either *Tie2-Cre* or *Pdgfb-CreER* transgenes disrupted the development of the pulmonary vasculature (Ren et al. 2014; Cai et al. 2016), indicating FOXF1 functions in an endothelial cell-autonomous manner during pulmonary angiogenesis. FOXF1 promotes VEGF, PDGF, and ANGPT/TIE2 signaling in the developing vasculature through direct transcriptional activation of *Flk1*, *Pdgfb*, and *Tek* genes.

1.5 Summary

Pulmonary vascular morphogenesis is mostly mediated by angiogenesis, instructed by paracrine signaling between endodermally derived epithelial progenitor cells and mesenchymal cells via processes mediated by VEGFs, ANGIOPOIETINs, PDGFs, FGFs, RETINOIC ACID, TGF-β, SHH, WNT, and NOTCH, which in turn control the activities of multiple transcription factors. Recent evidence supports important roles for the FOX, SOX, GATA, ETS, PRX, HIF, and PROX families of transcription factors in pulmonary vascular development. From a clinical perspective, FOXF1 transcription factor serves as a critical regulator of the pulmonary capillary network as haploinsufficiency of the *Foxf1* gene in mice and humans causes Alveolar Capillary Dysplasia. Mutations in the *FOXF1* locus lead to respiratory insufficiency in preterm infants associated with severe abnormalities in formation of the pulmonary vascular network. Better understanding of the signaling and transcriptional regulatory networks controlling development of the pulmonary vasculature will identify new therapeutic targets needed to treat congenital and acquired pulmonary disorders.

Acknowledgements We would like to thank Ann Maher for excellent editorial assistance. These studies were supported by NIH grants HL84151 and HL123490 (both to V.V.K.), HL132849 (to T.V.K.), and HL110964 and HL122642 (both to J.A.W.).

References

Aitola M, Carlsson P, Mahlapuu M, Enerback S, Pelto-Huikko M (2000) Forkhead transcription factor FoxF2 is expressed in mesodermal tissues involved in epithelio-mesenchymal interactions. Dev Dyn 218:136–149. https://doi.org/10.1002/(SICI)1097-0177(200005)218:1<136::AID-DVDY12>3.0.CO;2-U

Arora R, Papaioannou VE (2012) The murine allantois: a model system for the study of blood vessel formation. Blood 120:2562–2572. https://doi.org/10.1182/blood-2012-03-390070

Aubin J, Lemieux M, Tremblay M, Berard J, Jeannotte L (1997) Early postnatal lethality in Hoxa-5 mutant mice is attributable to respiratory tract defects. Dev Biol 192:432–445. https://doi.org/10.1006/dbio.1997.8746

Balli D, Ren X, Chou FS, Cross E, Zhang Y, Kalinichenko VV, Kalin TV (2012) Foxm1 transcription factor is required for macrophage migration during lung inflammation and tumor formation. Oncogene 31:3875–3888. https://doi.org/10.1038/onc.2011.549

Birdsey GM, Shah AV, Dufton N et al (2015) The endothelial transcription factor ERG promotes vascular stability and growth through Wnt/beta-catenin signaling. Dev Cell 32:82–96. https://doi.org/10.1016/j.devcel.2014.11.016

Bishop NB, Stankiewicz P, Steinhorn RH (2011) Alveolar capillary dysplasia. Am J Respir Crit Care Med 184:172–179. https://doi.org/10.1164/rccm.201010-1697CI

Bolte C, Zhang Y, Wang IC, Kalin TV, Molkentin JD, Kalinichenko VV (2011) Expression of Foxm1 transcription factor in cardiomyocytes is required for myocardial development. PLoS One 6:e22217. https://doi.org/10.1371/journal.pone.0022217

Bolte C, Zhang Y, York A, Kalin TV, Schultz Jel J, Molkentin JD, Kalinichenko VV (2012) Postnatal ablation of Foxm1 from cardiomyocytes causes late onset cardiac hypertrophy and fibrosis without exacerbating pressure overload-induced cardiac remodeling. PLoS One 7: e48713. https://doi.org/10.1371/journal.pone.0048713

Bolte C, Ren X, Tomley T et al (2015) Forkhead box F2 regulation of platelet-derived growth factor and myocardin/serum response factor signaling is essential for intestinal development. J Biol Chem 290:7563–7575. https://doi.org/10.1074/jbc.M114.609487

Bridges JP, Weaver TE (2006) Use of transgenic mice to study lung morphogenesis and function. ILAR J 47:22–31

Cai Y, Bolte C, Le T, Goda C, Xu Y, Kalin TV, Kalinichenko VV (2016) FOXF1 maintains endothelial barrier function and prevents edema after lung injury. Sci Signal 9:ra40. https://doi.org/10.1126/scisignal.aad1899

Cardoso WV, Lu J (2006) Regulation of early lung morphogenesis: questions, facts and controversies. Development 133:1611–1624. https://doi.org/10.1242/dev.02310

Carmeliet P, Mackman N, Moons L et al (1996) Role of tissue factor in embryonic blood vessel development. Nature 383:73–75. https://doi.org/10.1038/383073a0

Casie Chetty S, Rost MS, Enriquez JR et al (2017) Vegf signaling promotes vascular endothelial differentiation by modulating etv2 expression. Dev Biol. https://doi.org/10.1016/j.ydbio.2017.03.005

Cheng XH, Black M, Ustiyan V et al (2014) SPDEF inhibits prostate carcinogenesis by disrupting a positive feedback loop in regulation of the Foxm1 oncogene. PLoS Genet 10:e1004656. https://doi.org/10.1371/journal.pgen.1004656

Chisaka O, Capecchi MR (1991) Regionally restricted developmental defects resulting from targeted disruption of the mouse homeobox gene hox-1.5. Nature 350:473–479. https://doi.org/10.1038/350473a0

Costa RH, Kalinichenko VV, Lim L (2001) Transcription factors in mouse lung development and function. Am J Physiol Lung Cell Mol Physiol 280:L823–L838

Crivellato E (2011) The role of angiogenic growth factors in organogenesis. Int J Dev Biol 55:365–375. https://doi.org/10.1387/ijdb.103214ec

Cushing L, Costinean S, Xu W et al (2015) Disruption of miR-29 leads to aberrant differentiation of smooth muscle cells selectively associated with distal lung vasculature. PLoS Genet 11: e1005238. https://doi.org/10.1371/journal.pgen.1005238
De Val S, Black BL (2009) Transcriptional control of endothelial cell development. Dev Cell 16:180–195. https://doi.org/10.1016/j.devcel.2009.01.014
Dellinger MT, Meadows SM, Wynne K, Cleaver O, Brekken RA (2013) Vascular endothelial growth factor receptor-2 promotes the development of the lymphatic vasculature. PLoS One 8: e74686. https://doi.org/10.1371/journal.pone.0074686
Desai TJ, Chen F, Lu J et al (2006) Distinct roles for retinoic acid receptors alpha and beta in early lung morphogenesis. Dev Biol 291:12–24. https://doi.org/10.1016/j.ydbio.2005.10.045
Dharmadhikari AV, Szafranski P, Kalinichenko VV, Stankiewicz P (2015) Genomic and epigenetic complexity of the FOXF1 locus in 16q24.1: implications for development and disease. Curr Genomics 16:107–116. https://doi.org/10.2174/1389202916666150122223252
Dharmadhikari AV, Sun JJ, Gogolewski K et al (2016) Lethal lung hypoplasia and vascular defects in mice with conditional Foxf1 overexpression. Biol Open 5:1595–1606. https://doi.org/10.1242/bio.019208
Domyan ET, Ferretti E, Throckmorton K, Mishina Y, Nicolis SK, Sun X (2011) Signaling through BMP receptors promotes respiratory identity in the foregut via repression of Sox2. Development 138:971–981. https://doi.org/10.1242/dev.053694
El-Hashash AH, Al Alam D, Turcatel G, Bellusci S, Warburton D (2011a) Eyes absent 1 (Eya1) is a critical coordinator of epithelial, mesenchymal and vascular morphogenesis in the mammalian lung. Dev Biol 350:112–126. https://doi.org/10.1016/j.ydbio.2010.11.022
El-Hashash AH, Al Alam D, Turcatel G, Rogers O, Li X, Bellusci S, Warburton D (2011b) Six1 transcription factor is critical for coordination of epithelial, mesenchymal and vascular morphogenesis in the mammalian lung. Dev Biol 353:242–258. https://doi.org/10.1016/j.ydbio.2011.02.031
Ferrara N (1996) Vascular endothelial growth factor. Eur J Cancer 32A:2413–2422
Fischer A, Schumacher N, Maier M, Sendtner M, Gessler M (2004) The Notch target genes Hey1 and Hey2 are required for embryonic vascular development. Genes Dev 18:901–911. https://doi.org/10.1101/gad.291004
Fong GH, Rossant J, Gertsenstein M, Breitman ML (1995) Role of the Flt-1 receptor tyrosine kinase in regulating the assembly of vascular endothelium. Nature 376:66–70. https://doi.org/10.1038/376066a0
Francois M, Caprini A, Hosking B et al (2008) Sox18 induces development of the lymphatic vasculature in mice. Nature 456:643–647. https://doi.org/10.1038/nature07391
Galambos C, deMello DE (2007) Molecular mechanisms of pulmonary vascular development. Pediatr Dev Pathol 10:1–17. https://doi.org/10.2350/06-06-0122.1
Gao F, Bian F, Ma X, Kalinichenko VV, Das SK (2015) Control of regional decidualization in implantation: role of FoxM1 downstream of Hoxa10 and cyclin D3. Sci Rep 5:13863. https://doi.org/10.1038/srep13863
Goss AM, Tian Y, Tsukiyama T et al (2009) Wnt2/2b and beta-catenin signaling are necessary and sufficient to specify lung progenitors in the foregut. Dev Cell 17:290–298. https://doi.org/10.1016/j.devcel.2009.06.005
Grindley JC, Bellusci S, Perkins D, Hogan BL (1997) Evidence for the involvement of the Gli gene family in embryonic mouse lung development. Dev Biol 188:337–348. https://doi.org/10.1006/dbio.1997.8644
Grzenda A, Shannon J, Fisher J, Arkovitz MS (2013) Timing and expression of the angiopoietin-1-Tie-2 pathway in murine lung development and congenital diaphragmatic hernia. Dis Model Mech 6:106–114. https://doi.org/10.1242/dmm.008821
Habermehl D, Parkitna JR, Kaden S, Brugger B, Wieland F, Grone HJ, Schutz G (2011) Glucocorticoid activity during lung maturation is essential in mesenchymal and less in alveolar epithelial cells. Mol Endocrinol 25:1280–1288. https://doi.org/10.1210/me.2009-0380

Herriges M, Morrisey EE (2014) Lung development: orchestrating the generation and regeneration of a complex organ. Development 141:502–513. https://doi.org/10.1242/dev.098186

Hoffmann AD, Yang XH, Burnicka-Turek O et al (2014) Foxf genes integrate tbx5 and hedgehog pathways in the second heart field for cardiac septation. PLoS Genet 10:e1004604. https://doi.org/10.1371/journal.pgen.1004604

Hoggatt AM, Kim JR, Ustiyan V, Ren X, Kalin TV, Kalinichenko VV, Herring BP (2013) The transcription factor Foxf1 binds to serum response factor and myocardin to regulate gene transcription in visceral smooth muscle cells. J Biol Chem 288:28477–28487. https://doi.org/10.1074/jbc.M113.478974

Ihida-Stansbury K, McKean DM, Gebb SA et al (2004) Paired-related homeobox gene Prx1 is required for pulmonary vascular development. Circ Res 94:1507–1514. https://doi.org/10.1161/01.RES.0000130656.72424.20

Ihida-Stansbury K, Ames J, Chokshi M et al (2015) Role played by Prx1-dependent extracellular matrix properties in vascular smooth muscle development in embryonic lungs. Pulm Circ 5:382–397. https://doi.org/10.1086/681272

Ikeda K, Shaw-White JR, Wert SE, Whitsett JA (1996) Hepatocyte nuclear factor 3 activates transcription of thyroid transcription factor 1 in respiratory epithelial cells. Mol Cell Biol 16:3626–3636

Jones PL (2003) Homeobox genes in pulmonary vascular development and disease. Trends Cardiovasc Med 13:336–345

Kalin TV, Meliton L, Meliton AY, Zhu X, Whitsett JA, Kalinichenko VV (2008) Pulmonary mastocytosis and enhanced lung inflammation in mice heterozygous null for the Foxf1 gene. Am J Respir Cell Mol Biol 39:390–399. https://doi.org/10.1165/rcmb.2008-0044OC

Kalinichenko VV, Kalin TV (2015) Is there potential to target FOXM1 for 'undruggable' lung cancers? Expert Opin Ther Targets 19:865–867. https://doi.org/10.1517/14728222.2015.1042366

Kalinichenko VV, Lim L, Stolz DB et al (2001) Defects in pulmonary vasculature and perinatal lung hemorrhage in mice heterozygous null for the Forkhead Box f1 transcription factor. Dev Biol 235:489–506. https://doi.org/10.1006/dbio.2001.0322

Kalinichenko VV, Zhou Y, Shin B, Stolz DB, Watkins SC, Whitsett JA, Costa RH (2002) Wild-type levels of the mouse Forkhead Box f1 gene are essential for lung repair. Am J Physiol Lung Cell Mol Physiol 282:L1253–L1265. https://doi.org/10.1152/ajplung.00463.2001

Kalinichenko VV, Bhattacharyya D, Zhou Y, Gusarova GA, Kim W, Shin B, Costa RH (2003a) Foxf1 +/- mice exhibit defective stellate cell activation and abnormal liver regeneration following CCl4 injury. Hepatology 37:107–117. https://doi.org/10.1053/jhep.2003.50005

Kalinichenko VV, Gusarova GA, Shin B, Costa RH (2003b) The forkhead box F1 transcription factor is expressed in brain and head mesenchyme during mouse embryonic development. Gene Expr Patterns 3:153–158

Kalinichenko VV, Gusarova GA, Kim IM et al (2004) Foxf1 haploinsufficiency reduces Notch-2 signaling during mouse lung development. Am J Physiol Lung Cell Mol Physiol 286:L521–L530. https://doi.org/10.1152/ajplung.00212.2003

Kanai-Azuma M, Kanai Y, Gad JM et al (2002) Depletion of definitive gut endoderm in Sox17-null mutant mice. Development 129:2367–2379

Kataoka H, Hayashi M, Nakagawa R et al (2011) Etv2/ER71 induces vascular mesoderm from Flk1+PDGFRalpha+ primitive mesoderm. Blood 118:6975–6986. https://doi.org/10.1182/blood-2011-05-352658

Kim IM, Ramakrishna S, Gusarova GA, Yoder HM, Costa RH, Kalinichenko VV (2005) The forkhead box m1 transcription factor is essential for embryonic development of pulmonary vasculature. J Biol Chem 280:22278–22286. https://doi.org/10.1074/jbc.M500936200

Kimura S, Hara Y, Pineau T, Fernandez-Salguero P, Fox CH, Ward JM, Gonzalez FJ (1996) The T/ebp null mouse: thyroid-specific enhancer-binding protein is essential for the organogenesis of the thyroid, lung, ventral forebrain, and pituitary. Genes Dev 10:60–69

Kimura Y, Suzuki T, Kaneko C et al (2002) Retinoid receptors in the developing human lung. Clin Sci (Lond) 103:613–621

Kool H, Mous D, Tibboel D, de Klein A, Rottier RJ (2014) Pulmonary vascular development goes awry in congenital lung abnormalities. Birth Defects Res C Embryo Today 102:343–358. https://doi.org/10.1002/bdrc.21085

Kumar VH, Lakshminrusimha S, El Abiad MT, Chess PR, Ryan RM (2005) Growth factors in lung development. Adv Clin Chem 40:261–316

Kume T, Jiang H, Topczewska JM, Hogan BL (2001) The murine winged helix transcription factors, Foxc1 and Foxc2, are both required for cardiovascular development and somitogenesis. Genes Dev 15:2470–2482. https://doi.org/10.1101/gad.907301

Lange AW, Haitchi HM, LeCras TD et al (2014) Sox17 is required for normal pulmonary vascular morphogenesis. Dev Biol 387:109–120. https://doi.org/10.1016/j.ydbio.2013.11.018

Lee D, Park C, Lee H et al (2008) ER71 acts downstream of BMP, Notch, and Wnt signaling in blood and vessel progenitor specification. Cell Stem Cell 2:497–507. https://doi.org/10.1016/j.stem.2008.03.008

Lee JM, Kwon HJ, Lai WF, Jung HS (2014a) Requirement of Runx3 in pulmonary vasculogenesis. Cell Tissue Res 356:445–449. https://doi.org/10.1007/s00441-014-1816-x

Lee SH, Lee S, Yang H et al (2014b) Notch pathway targets proangiogenic regulator Sox17 to restrict angiogenesis. Circ Res 115:215–226. https://doi.org/10.1161/CIRCRESAHA.115.303142

Lelievre E, Lionneton F, Soncin F, Vandenbunder B (2001) The Ets family contains transcriptional activators and repressors involved in angiogenesis. Int J Biochem Cell Biol 33:391–407

Li A, Hardy R, Stoner S, Tuckermann J, Seibel M, Zhou H (2013) Deletion of mesenchymal glucocorticoid receptor attenuates embryonic lung development and abdominal wall closure. PLoS One 8:e63578. https://doi.org/10.1371/journal.pone.0063578

Liu Y, Sadikot RT, Adami GR et al (2011) FoxM1 mediates the progenitor function of type II epithelial cells in repairing alveolar injury induced by Pseudomonas aeruginosa. J Exp Med 208:1473–1484. https://doi.org/10.1084/jem.20102041

Lu Y, Thomson JM, Wong HY, Hammond SM, Hogan BL (2007) Transgenic over-expression of the microRNA miR-17-92 cluster promotes proliferation and inhibits differentiation of lung epithelial progenitor cells. Dev Biol 310:442–453. https://doi.org/10.1016/j.ydbio.2007.08.007

Lufkin T, Dierich A, LeMeur M, Mark M, Chambon P (1991) Disruption of the Hox-1.6 homeobox gene results in defects in a region corresponding to its rostral domain of expression. Cell 66:1105–1119

Lyden D, Young AZ, Zagzag D et al (1999) Id1 and Id3 are required for neurogenesis, angiogenesis and vascularization of tumour xenografts. Nature 401:670–677. https://doi.org/10.1038/44334

Madison BB, McKenna LB, Dolson D, Epstein DJ, Kaestner KH (2009) FoxF1 and FoxL1 link hedgehog signaling and the control of epithelial proliferation in the developing stomach and intestine. J Biol Chem 284:5936–5944. https://doi.org/10.1074/jbc.M808103200

Mahlapuu M, Enerback S, Carlsson P (2001) Haploinsufficiency of the forkhead gene Foxf1, a target for sonic hedgehog signaling, causes lung and foregut malformations. Development 128:2397–2406

Malin D, Kim IM, Boetticher E et al (2007) Forkhead box F1 is essential for migration of mesenchymal cells and directly induces integrin-beta3 expression. Mol Cell Biol 27:2486–2498. https://doi.org/10.1128/MCB.01736-06

Marcelo KL, Goldie LC, Hirschi KK (2013) Regulation of endothelial cell differentiation and specification. Circ Res 112:1272–1287. https://doi.org/10.1161/CIRCRESAHA.113.300506

McGowan S, Jackson SK, Jenkins-Moore M, Dai HH, Chambon P, Snyder JM (2000) Mice bearing deletions of retinoic acid receptors demonstrate reduced lung elastin and alveolar numbers. Am J Respir Cell Mol Biol 23:162–167. https://doi.org/10.1165/ajrcmb.23.2.3904

Metzger RJ, Klein OD, Martin GR, Krasnow MA (2008) The branching programme of mouse lung development. Nature 453:745–750. https://doi.org/10.1038/nature07005

Milewski D, Pradhan A, Wang X et al (2017) FoxF1 and FoxF2 transcription factors synergistically promote rhabdomyosarcoma carcinogenesis by repressing transcription of p21Cip1 CDK inhibitor. Oncogene 36:850–862. https://doi.org/10.1038/onc.2016.254

Miniati D, Jelin EB, Ng J et al (2010) Constitutively active endothelial Notch4 causes lung arteriovenous shunts in mice. Am J Physiol Lung Cell Mol Physiol 298:L169–L177. https://doi.org/10.1152/ajplung.00188.2009

Morrisey EE, Hogan BL (2010) Preparing for the first breath: genetic and cellular mechanisms in lung development. Dev Cell 18:8–23. https://doi.org/10.1016/j.devcel.2009.12.010

Motoyama J, Liu J, Mo R, Ding Q, Post M, Hui CC (1998) Essential function of Gli2 and Gli3 in the formation of lung, trachea and oesophagus. Nat Genet 20:54–57. https://doi.org/10.1038/1711

Niederreither K, Subbarayan V, Dolle P, Chambon P (1999) Embryonic retinoic acid synthesis is essential for early mouse post-implantation development. Nat Genet 21:444–448. https://doi.org/10.1038/7788

Park HL, Bai C, Platt KA et al (2000) Mouse Gli1 mutants are viable but have defects in SHH signaling in combination with a Gli2 mutation. Development 127:1593–1605

Park C, Kim TM, Malik AB (2013) Transcriptional regulation of endothelial cell and vascular development. Circ Res 112:1380–1400. https://doi.org/10.1161/CIRCRESAHA.113.301078

Pereira FA, Qiu Y, Zhou G, Tsai MJ, Tsai SY (1999) The orphan nuclear receptor COUP-TFII is required for angiogenesis and heart development. Genes Dev 13:1037–1049

Peterson RS, Lim L, Ye H, Zhou H, Overdier DG, Costa RH (1997) The winged helix transcriptional activator HFH-8 is expressed in the mesoderm of the primitive streak stage of mouse embryos and its cellular derivatives. Mech Dev 69:53–69

Petrova TV, Karpanen T, Norrmen C et al (2004) Defective valves and abnormal mural cell recruitment underlie lymphatic vascular failure in lymphedema distichiasis. Nat Med 10:974–981. https://doi.org/10.1038/nm1094

Pradhan A, Ustiyan V, Zhang Y, Kalin TV, Kalinichenko VV (2016) Forkhead transcription factor FoxF1 interacts with Fanconi anemia protein complexes to promote DNA damage response. Oncotarget 7:1912–1926. 10.18632/oncotarget.6422

Ren X, Zhang Y, Snyder J, Cross ER, Shah TA, Kalin TV, Kalinichenko VV (2010) Forkhead box M1 transcription factor is required for macrophage recruitment during liver repair. Mol Cell Biol 30:5381–5393. https://doi.org/10.1128/MCB.00876-10

Ren X, Shah TA, Ustiyan V et al (2013) FOXM1 promotes allergen-induced goblet cell metaplasia and pulmonary inflammation. Mol Cell Biol 33:371–386. https://doi.org/10.1128/MCB.00934-12

Ren X, Ustiyan V, Pradhan A et al (2014) FOXF1 transcription factor is required for formation of embryonic vasculature by regulating VEGF signaling in endothelial cells. Circ Res 115:709–720. https://doi.org/10.1161/CIRCRESAHA.115.304382

Sakamoto Y, Hara K, Kanai-Azuma M et al (2007) Redundant roles of Sox17 and Sox18 in early cardiovascular development of mouse embryos. Biochem Biophys Res Commun 360:539–544. https://doi.org/10.1016/j.bbrc.2007.06.093

Sauvageau M, Goff LA, Lodato S et al (2013) Multiple knockout mouse models reveal lincRNAs are required for life and brain development. Elife 2:e01749. https://doi.org/10.7554/eLife.01749

Sen P, Dharmadhikari AV, Majewski T et al (2014) Comparative analyses of lung transcriptomes in patients with alveolar capillary dysplasia with misalignment of pulmonary veins and in foxf1 heterozygous knockout mice. PLoS One 9:e94390. https://doi.org/10.1371/journal.pone.0094390

Sengupta A, Kalinichenko VV, Yutzey KE (2013) FoxO1 and FoxM1 transcription factors have antagonistic functions in neonatal cardiomyocyte cell-cycle withdrawal and IGF1 gene regulation. Circ Res 112:267–277. https://doi.org/10.1161/CIRCRESAHA.112.277442

Shalaby F, Rossant J, Yamaguchi TP, Gertsenstein M, Wu XF, Breitman ML, Schuh AC (1995) Failure of blood-island formation and vasculogenesis in Flk-1-deficient mice. Nature 376:62–66. https://doi.org/10.1038/376062a0

Shaw-White JR, Bruno MD, Whitsett JA (1999) GATA-6 activates transcription of thyroid transcription factor-1. J Biol Chem 274:2658–2664

Spyropoulos DD, Pharr PN, Lavenburg KR, Jackers P, Papas TS, Ogawa M, Watson DK (2000) Hemorrhage, impaired hematopoiesis, and lethality in mouse embryos carrying a targeted disruption of the Fli1 transcription factor. Mol Cell Biol 20:5643–5652

Stankiewicz P, Sen P, Bhatt SS et al (2009) Genomic and genic deletions of the FOX gene cluster on 16q24.1 and inactivating mutations of FOXF1 cause alveolar capillary dysplasia and other malformations. Am J Hum Genet 84:780–791. https://doi.org/10.1016/j.ajhg.2009.05.005

Szafranski P, Dharmadhikari AV, Brosens E et al (2013) Small noncoding differentially methylated copy-number variants, including lncRNA genes, cause a lethal lung developmental disorder. Genome Res 23:23–33. https://doi.org/10.1101/gr.141887.112

Tibboel J, Groenman FA, Selvaratnam J et al (2015) Hypoxia-inducible factor-1 stimulates postnatal lung development but does not prevent O2-induced alveolar injury. Am J Respir Cell Mol Biol 52:448–458. https://doi.org/10.1165/rcmb.2014-0037OC

Tiozzo C, Carraro G, Al Alam D et al (2012) Mesodermal Pten inactivation leads to alveolar capillary dysplasia-like phenotype. J Clin Invest 122:3862–3872. https://doi.org/10.1172/JCI61334

Ustiyan V, Wang IC, Ren X et al (2009) Forkhead box M1 transcriptional factor is required for smooth muscle cells during embryonic development of blood vessels and esophagus. Dev Biol 336:266–279. https://doi.org/10.1016/j.ydbio.2009.10.007

Ustiyan V, Wert SE, Ikegami M, Wang IC, Kalin TV, Whitsett JA, Kalinichenko VV (2012) Foxm1 transcription factor is critical for proliferation and differentiation of Clara cells during development of conducting airways. Dev Biol 370:198–212. https://doi.org/10.1016/j.ydbio.2012.07.028

Ustiyan V, Zhang Y, Perl AK, Whitsett JA, Kalin TV, Kalinichenko VV (2016) beta-Catenin and Kras/Foxm1 signaling pathway are critical to restrict Sox9 in basal cells during pulmonary branching morphogenesis. Dev Dyn 245:590–604. https://doi.org/10.1002/dvdy.24393

van Tuyl M, Liu J, Groenman F et al (2006) Iroquois genes influence proximo-distal morphogenesis during rat lung development. Am J Physiol Lung Cell Mol Physiol 290:L777–L789. https://doi.org/10.1152/ajplung.00293.2005

van Tuyl M, Groenman F, Wang J, Kuliszewski M, Liu J, Tibboel D, Post M (2007) Angiogenic factors stimulate tubular branching morphogenesis of sonic hedgehog-deficient lungs. Dev Biol 303:514–526. https://doi.org/10.1016/j.ydbio.2006.11.029

Ventura A, Young AG, Winslow MM et al (2008) Targeted deletion reveals essential and overlapping functions of the miR-17 through 92 family of miRNA clusters. Cell 132:875–886. https://doi.org/10.1016/j.cell.2008.02.019

Wang X, Bhattacharyya D, Dennewitz MB, Kalinichenko VV, Zhou Y, Lepe R, Costa RH (2003) Rapid hepatocyte nuclear translocation of the Forkhead Box M1B (FoxM1B) transcription factor caused a transient increase in size of regenerating transgenic hepatocytes. Gene Expr 11:149–162

Wang IC, Zhang Y, Snyder J et al (2010) Increased expression of FoxM1 transcription factor in respiratory epithelium inhibits lung sacculation and causes Clara cell hyperplasia. Dev Biol 347:301–314. https://doi.org/10.1016/j.ydbio.2010.08.027

Wang IC, Snyder J, Zhang Y et al (2012) Foxm1 mediates cross talk between Kras/mitogen-activated protein kinase and canonical Wnt pathways during development of respiratory epithelium. Mol Cell Biol 32:3838–3850. https://doi.org/10.1128/MCB.00355-12

Wang IC, Ustiyan V, Zhang Y, Cai Y, Kalin TV, Kalinichenko VV (2014) Foxm1 transcription factor is required for the initiation of lung tumorigenesis by oncogenic Kras(G12D). Oncogene 33:5391–5396. https://doi.org/10.1038/onc.2013.475

Warburton D, Schwarz M, Tefft D, Flores-Delgado G, Anderson KD, Cardoso WV (2000) The molecular basis of lung morphogenesis. Mech Dev 92:55–81

Warburton D, El-Hashash A, Carraro G et al (2010) Lung organogenesis. Curr Top Dev Biol 90:73–158. https://doi.org/10.1016/S0070-2153(10)90003-3

White AC, Xu J, Yin Y, Smith C, Schmid G, Ornitz DM (2006) FGF9 and SHH signaling coordinate lung growth and development through regulation of distinct mesenchymal domains. Development 133:1507–1517. https://doi.org/10.1242/dev.02313

Wigle JT, Oliver G (1999) Prox1 function is required for the development of the murine lymphatic system. Cell 98:769–778

Wilhelm K, Happel K, Eelen G et al (2016) FOXO1 couples metabolic activity and growth state in the vascular endothelium. Nature 529:216–220. https://doi.org/10.1038/nature16498

Xia H, Ren X, Bolte CS et al (2015) Foxm1 regulates resolution of hyperoxic lung injury in newborns. Am J Respir Cell Mol Biol 52:611–621. https://doi.org/10.1165/rcmb.2014-0091OC

Xu J, Liu H, Lan Y, Aronow BJ, Kalinichenko VV, Jiang R (2016) A Shh-Foxf-Fgf18-Shh molecular circuit regulating palate development. PLoS Genet 12:e1005769. https://doi.org/10.1371/journal.pgen.1005769

Yun EJ, Lorizio W, Seedorf G, Abman SH, Vu TH (2016) VEGF and endothelium-derived retinoic acid regulate lung vascular and alveolar development. Am J Physiol Lung Cell Mol Physiol 310:L287–L298. https://doi.org/10.1152/ajplung.00229.2015

Zeng X, Wert SE, Federici R, Peters KG, Whitsett JA (1998) VEGF enhances pulmonary vasculogenesis and disrupts lung morphogenesis in vivo. Dev Dyn 211:215–227. https://doi.org/10.1002/(SICI)1097-0177(199803)211:3<215::AID-AJA3>3.0.CO;2-K

Zeng X, Gray M, Stahlman MT, Whitsett JA (2001) TGF-beta1 perturbs vascular development and inhibits epithelial differentiation in fetal lung in vivo. Dev Dyn 221:289–301. https://doi.org/10.1002/dvdy.1140

Chapter 2
Comparative View of Lung Vascular Endothelium of Cattle, Horses, and Water Buffalo

David Schneberger, Ram S. Sethi, and Baljit Singh

Abstract Endothelium plays an important role in maintaining the vascular barrier and physiological homeostasis. Endothelium also is fundamental to the initiation and regulation of inflammation. Endothelium demonstrates phenotypic and functional heterogeneity not only among various organs but also within an organ. One of the striking examples would be the pulmonary endothelium that participates in creating blood–air barrier. Endothelium in large pulmonary blood vessels is distinct in structure and function from that lining of the pulmonary capillaries. This chapter focuses on the comparative aspects of pulmonary endothelium and highlight unique differences such as the presence of pulmonary intravascular macrophages among select species.

2.1 Introduction

The pulmonary endothelium is a dynamic and metabolically active layer of squamous endothelial cells that is ideally placed to mediate lung homoeostasis (Millar et al. 2016). The pulmonary capillary is made up of a few (usually 2–3) thin, squamous endothelial cells (Dornan and Meban 1985; King et al. 2004), and capillary endothelial cells constitute 30–50% and 46–50% of the total cell population in the alveolar septal wall

D. Schneberger
Canadian Centre for Health and Safety in Agriculture, University of Saskatchewan, Saskatoon, SK, Canada
e-mail: david.schneberger@usask.ca

R.S. Sethi
School of Animal Biotechnology, Guru Angad Dev Veterinary and Animal Sciences University, Ludhiana, India
e-mail: rs.sethi@usask.ca

B. Singh (✉)
Faculty of Veterinary Medicine, University of Calgary, Calgary, AB, Canada
e-mail: baljit.singh1@ucalgary.ca

K. Parthasarathi (ed.), *Molecular and Functional Insights Into the Pulmonary Vasculature*, Advances in Anatomy, Embryology and Cell Biology 228,
https://doi.org/10.1007/978-3-319-68483-3_2

in human (Crapo et al. 1982) and nonhuman primate (Crapo et al. 1983), respectively. Horse lung showed an average of 252.4 capillaries per centimeter of lung alveolar wall with endothelial cells constituting 49.5% of total tissue cells in the alveolar septal wall (Gillespie and Tyler 1967). Endothelium is the gatekeeper of the tissues and regulates traffic of circulating cells and molecules into the tissues (Cotran 1989).

Endothelial cells of the lung can be easily split into two subgroups, one comprising the lining of larger blood vessels such as arteries and veins and another population that constitutes capillary endothelial cells of the microvasculature, each with distinct functions. Both subsets of endothelial cells contain rough endoplasmic reticulum, numerous vesicles such as caveolae, a low number of mitochondria, and filaments around the basal cell surface amongst other common cellular features (King et al. 2004). Filaments are more common in arterial cells which also exhibit projections from the apical surface, particularly near cellular junctions (King et al. 2004). As there is transition from arterial to microvascular endothelial cells, there is a flattening and elongation of the cells, with a greater restriction of cytoplasm near the flattened periphery. Many of the organelles are excluded from the flattened parts and concentrated around the nucleus. While apical projections were present, they were not as pronounced in this population (King et al. 2004; Ochoa et al. 2010).

A key intracellular feature of endothelial cells are dark bodies in the cytoplasm referred to as Weibel–Palade bodies (WPb) which contain a variety of secretable products such as von Willebrand Factor (vWF), interleukin (IL)-8, factor XIIIa, and P-selectin. While these bodies are present in arterial endothelial cells they are not present in the capillary endothelial cells, though these factors are still secreted by these cells as well (Fuchs and Weibel 1966; Lowenstein et al. 2005). Upon injury or another activating stimulus, these bodies fuse with the cellular membrane and degranulate, releasing their contents into circulation and on the surface of the endothelial cells. vWF, a substrate of XIIIa, plays a role in adhesion of platelets to the endothelium and also serves as a carrier for coagulation factor VIII (functions reviewed in Wang and Eikenboom 2010). IL-8 on the other hand is a strong neutrophil chemoattractant and activator, which is likely to cause damage to the endothelium (discussed later; Fig. 2.1). vWF can bind to platelets and initiate clot formation (Ochoa et al. 2010). While this is one of the most obvious and recognizable difference between both endothelial cell populations, a host of other differences exist.

Another function of the endothelium is the active regulation of vascular permeability and relaxation of arterial vessels. Relaxation is controlled by nitric oxide (NO) release from arterial endothelial cells (Furchgott and Vanhoutte 1989; Furchgott and Zawadzki 1980) which in turn prompts smooth muscle to release guanylate cyclase, increasing cGMP and vasodilation (Furchgott and Vanhoutte 1989). This mechanism is observed during exercise (Pelletier and Leith 1993), even at rest (Rees et al. 1989), and can be prompted by shear force of circulating blood (Kaiser and Sparks 1986; Koller et al. 1994). In hypoxia, removal of endothelial cells from vessels significantly reduces arterial NO production and the resulting vasoconstriction (around 65–70%), indicating a role for these cells in this phenomenon as well (MacEachern et al. 2004). However, such removal in human and sheep completely abrogated this response (Demiryurek et al. 1991, 1993), suggesting that this function is somewhat less

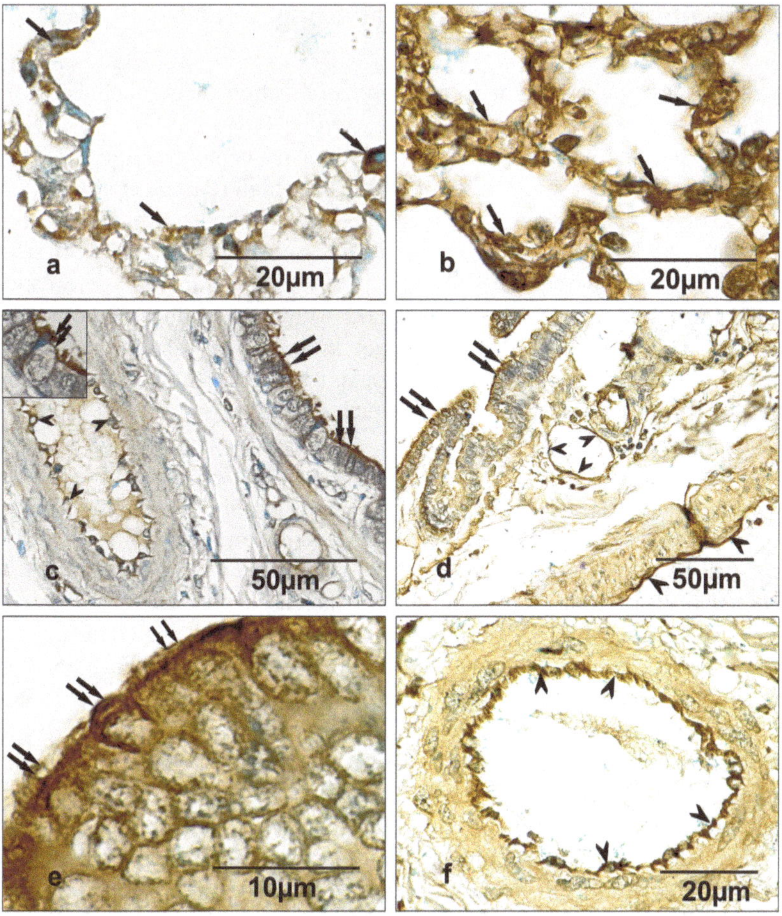

Fig. 2.1 (a) IL-8 expression (arrows) is seen in the alveolar septa of normal and (b) inflamed lungs. (c and e) The airway epithelium (double arrows) and vascular endothelium (arrowheads) of the normal lungs reacts weakly for IL-8 compared with (d) the intense expression in the airway epithelium (double arrows) and (f) the vascular endothelium (arrowheads) of the inflamed lungs. IHC. Reprinted with permission from J Comp Pathol. 2011 Feb–Apr;144(2–3):135–44. doi: https://doi.org/10.1016/j.jcpa.2010.08.003

important in horses and cattle than in other species (MacEachern et al. 2004). The role of NO in this phenomenon may be more complex though as in exercise models of lung injury, inhibition of NO production in horses increased damage seen in the lungs, suggesting mechanisms other than just vessel relaxation (pressure) likely play a role in endothelial damage (Kindig et al. 2000).

2.2 Circulation Constraints in the Lung

Maintenance of vascular integrity is an important function of endothelial cells in the lung. They can however often fail during exertion (Pascoe et al. 1981). Several studies have focused on exercise induced pulmonary hemorrhage in the lungs of race horses, where blood pressure can become very high (Pascoe et al. 1981; West et al. 1993). Arterial pressures can reach up to 120 mmHg in exercising horses (West et al. 1993), and capillary threshold pressure can reach between 75 and 100 mmHg in thoroughbreds (Birks et al. 1997). By comparison, the capillary threshold in rabbits is below 40 mmHg, and 70 mmHg in dogs (Fu et al. 1992; Tsukimoto et al. 1991). While we do not know of any study that examines molecular structural differences between rabbit and horse lung that can account for these differences, it seems likely that such differences must exist to allow for such increased pressures in the capillaries. A study estimating canine and equine capillary strength based on morphology, while accurately predicting increased strength over rabbits, also underestimated actual strength in equines (Birks et al. 1994), suggesting such molecular differences should be present.

Response to pressure in endothelial cells is carried out by the calcium channel TRPV4 (Jian et al. 2008; Yin et al. 2008). Increased pressure elevates cellular calcium, but blockage of TRPV can abrogate most of this change (Jian et al. 2008; Yin et al. 2008). Activation of the channel can also increase lung endothelial permeability (Alvarez et al. 2006). Increase in pressure is also shown to induce exocytosis of endothelial vesicles such as WPb (Kuebler et al. 1999). As these bodies include proteins such as vWF, P-selectin, and IL-8, pressure and calcium control could also be considered mechanisms of inflammation in these cells. Even eNOS production in these cells is controlled partially by calcium and can act as a negative feedback on TRPV4 induced calcium influx (Yin et al. 2008).

Finally, endothelial cells are constrained by the competing requirements of gas diffusion and blood pressure. Thus, it is no surprise that their capillary cell morphology takes on a more flattened structure compared to arterial endothelial cells to improve gas exchange. Increased ability to handle pressure will come at a cost of increased thickness, which will also trade off gas exchange efficiency. Thoroughbred horses in particular fall into this conundrum. While bred for work and speed, cardiac output has been increased, but the lungs often cannot handle this output pressure, such that alveolar bleeding is a common occurrence (West 2000). In the end, selection pressure may remain on improving endothelial cell thinness to maintain gas diffusion efficiency, and increase of extracelluar matrix toughness to improve strength of the alveoli (West 2000). While improvement of endothelial cell junctions would be expected, disruptions when they happen are observed frequently in the cells and not at junctions (Costello et al. 1992), and often seal quickly upon relief of pressure (Elliott et al. 1992).

Capillary endothelium is surrounded by mainly alveolar type I epithelial cells and occasional type II cell (Townsley 2012). Further, capillary endothelium is connected to other cellular constituents of the alveolar septal wall via fibroblasts by intercellular junctions (Sirianni et al. 2003) and pericytes by myoendothelial junctions (Michel et al. 1995). Most of these cell junctions are not random but strategically placed at the interface between the thick and the thin sides of the septal wall to provide structural stability and minimize the distensibility, especially during the increase in the mass of surrounding tissue (Walker et al. 1995).

Capillary endothelium lacks large gap junctions unlike endothelial junctions in extra-alveolar arteries and veins, but contains occasional small gaps in capillary inter-endothelial cell junctional complexes (Schneeberger 1982). Further, a third adjoining endothelial cell forms a flap overlaying the borders of these cells and this flap has specific organization of junctional strands (Walker et al. 1994). These strands run parallel instead of perpendicular to the plane of the overlying cell to provide some protection against transendothelial fluid movement during increased intravascular pressure as the increased hydrostatic force results in "sealing" of this junctional complex (Walker et al. 1994).

Endothelial cells are capable of allowing diffusion of a number of molecules across the cellular barrier, but studies in several species have shown a general resistance to diffusion, particularly in lung capillary endothelial cells. Capillary endothelial cells of the lung show lower permeability to water and solutes compared to vascular endothelial cells (Kelly et al. 1998; Parker and Yoshikawa 2002; Parker et al. 2006). Studies of hydraulic conductance also show values much lower for microvascular endothelial cells suggesting tighter junctions and lower diffusion (Kelly et al. 1998; Parker et al. 2006; Ofori-Acquah et al. 2008). This makes sense when considering the need to maximize diffusion of oxygen across the cell membrane and minimize solute and fluid transport, the last of which could cause tissue or alveolar edema problems if not closely controlled. Both arterial and microvascular endothelial cells have about the same amount of intracellular actin. However, the chemical disruption of actin wtih cytochalasin D is much harder in microvascular endothelial cells (Ofori-Acquah et al. 2008).

A reason for this more restrictive permeability is alterations in localized cAMP expression in endothelial cells. High levels of cAMP are generated at the membrane of microvascular endothelial cells by adenylate cyclase 6 which enhances barrier function (Ludwig and Seuwen 2002). This adenylate cyclase is calcium regulated and thus permeability can be increased by increased intracellular (cytosolic) calcium (Cioffi et al. 2002; Stevens et al. 1995; Sayner and Stevens 2006; Sayner et al. 2006), though a specific cAMP effector of barrier strength is not fully resolved (Prasain and Stevens 2009). One such example is endothelial response to thrombin. Thrombin is a pro-inflammatory molecule that binds to PAR receptors (PAR-1 and PAR-3) and activates endothelial cells (reviewed in Minami et al. 2004). Amongst these cellular changes is Gq activation, which leads to calcium release from the endoplasmic reticulum resulting in calcium depletion which is replenished by opening calcium channels (TRPC1 and TRPC4 subunits). This increase in free cytosolic calcium thus disrupts the endothelial cell barrier (Cioffi et al. 2009). Another ion channel, TVRP4, is selectively expressed in lung microvasculature but not the arterial endothelial cells (Alvarez et al. 2006). Activation of this channel leads to leakage into the alveoli, but in this case leakage is due to sloughing of cells and/or loss of attachment to cell matrix (Alvarez et al. 2006).

Neuropeptides play important roles in the regulation of respiratory function and topographical inflammation and are localized in both motor and sensory neurons in the mammalian respiratory tract (Lundberg et al. 1988). Substance P (SP) and calcitonin gene-related peptide (CGRP)-immunoreactive nerve fibers are present around pulmonary blood vessels throughout the respiratory tract of calves and cows (Nishi et al. 2000). These fibers are few in number in the lung compared to nasal and laryngeal mucosae and tracheal bronchus, and are more numerous in calves than in cows (Nishi et al. 2000). Co-localization of SP and CGRP in most of the nerve fibers suggests that these nerve fibers are involved in the regulation of the bovine respiratory tract.

2.3 Inflammation and Immunity

Endothelial cells are located at the interface between blood and tissues not only to gate the traffic of molecules and cells across the vessel wall but also contribute to hemostasis, inflammatory reactions, and immunity (Mantovani et al. 1992). When not activated, endothelial cells can secrete a variety of products that can aid in inhibiting clotting and inappropriate activation of innate immune functions (reviewed in Pober and Sessa 2007). Amongst these is NO, which can inhibit platelet aggregation and adhesion in addition to being a vasorelaxant as mentioned earlier (Sessa 2004). In this state, P-selectin is sequestered within the cell (Bonfanti et al. 1989), and other leukocytes adherence molecules such as E-selectin, VCAM, and ICAM-1 are suppressed to a greater degree (Pober and Sessa 2007).

During inflammation, there is binding to cell-surface receptors by one of a number of potential molecules (e.g., histamine, thrombin, LPS, etc.). For those that bind to G-protein receptors, there is a release of cytosolic Ca^{2+} that will result in increased endothelial permeability as discussed earlier (Pober and Cotran 1990; Pober and Sessa 2007) causing release of WPb (Birch et al. 1992, 1994). Calcium also activates phospholipase A2 which catalyzes arachidonic acid formation and subsequently its conversion to COX1 resulting in increased blood flow (Egan and FitzGerald 2006). A by-product of arachidonic acid formation, lysophosphatidylcholine, at the same time acts as a platelet activating factor. When this is combined with the released P-selectin it leads to neutrophil attachment, integrin activation, and the extravasation of said neutrophils across the endothelial barrier (Prescott et al. 1984; Lorant et al. 1991; Pober and Sessa 2007).

Another activated protein, RHO, also aids in contraction of cellular actin filaments, resulting in creation of gaps between endothelial cells, especially post-capillary venules (Heltianu et al. 1982). Passage of most cells through the endothelial barrier appears to be between such endothelial cells and supported by increased expression in these intercellular gaps of PECAM1 and CD99 (Marchesi 1961; Schenkel et al. 2002). Cytokines such as TNF-α and IL-1 can also induce a similar response (reviewed in Pober and Sessa 2007), but leading to transcriptional activation of NF-kB and AP-1, which activates a number of the same effector pathways already mentioned (Pober and Sessa 2007). Secretion of IL-8 will lead to activation, firm attachment, and diapedesis of neutrophils to the endothelial cells, but still requires E-selectin to tether such cells (Pan et al. 1998).

Specific to pulmonary microvascular capillary endothelial cells is a 1G T-type calcium channel. Activated by thrombin, the channel causes membrane depolarization, but in this case does not increase permeability but instead promotes P-selectin expression for neutrophil recruitment (Wu et al. 2003, 2009). Thus, calcium again appears to play a very important role in induction of inflammatory responses in endothelial cells, particularly with regard to P-selectin expression (Kuebler et al. 1999; Parthasarathi et al. 2006).

2.4 Toll-Like Receptors

Toll-like receptors (TLRs), the mammalian homologues of the Drosophila Toll family, are critical for recognition of conserved pathogen-associated molecular patterns in bacteria and viruses by immune cells (Aderem 2001). TLRs are expressed in the endothelium to activate the immune system when encountering specific antigen (Xu et al. 2011). TLR4 recognizes lipopolysaccharides (LPS) and plays important roles in host defense against bacterial infections (Lymboussaki et al. 1998; Takeda et al. 2003). We reported the presence of weak TLR4 immunoreactivity in the vascular endothelium in normal lung of cattle (Wassef et al. 2004) and buffalo (Sethi et al. 2011), which is consistent to what is observed in humans (Faure et al. 2000). Some TLR4 staining was reported in peribronchiolar blood vessels and alveolar septal endothelial cells in horse lungs (Singh Suri et al. 2006). TLR4 activates endothelium and induces recruitment of leukocytes in lung along with expression of adhesion molecules such as P-selectin and vascular cell adhesion molecule-1 (Andonegui et al. 2002, 2003). Interestingly, TLR4 appeared unchanged or reduced in the vascular endothelium of inflamed lungs from calves and Water buffalo infected with *Mannheimia haemolytica* and *Pasteurella multocida*, respectively (Wassef et al. 2004; Sethi et al. 2011). Lack of TLR4 in vascular endothelium during *Mannheimia haemolytica* and *Pasteurella multocida* infection may be a protective mechanism against unwanted inflammation.

Similarly, early studies with LPS treatment of horse lungs showed endothelial cells to be fairly unaffected by the exposure. This was, however, in the absence of immune cells such as neutrophils. While this study primarily looked at electron microscopic examination of endothelial cells, it does suggest that endothelial cells maintain morphology through exposure. By contrast, cell death was common in the presence of neutrophils, suggesting that when other immune cells are present endothelial cells are sensitive to products of these cells in the horse (Turek et al. 1987). Later experiments have also suggested TNF-α and IL-1b secretion by these cells to LPS in a whole animal exposure system (Parbhakar et al. 2005).

TLR9 identifies bacterial DNA and is a key player in bacterial DNA induced cell signaling (Bauer and Wagner 2002). TLR9 is not only activated by unmethylated CpG motifs within ssDNA, but also by CpG motifs in nucleic acids released during vascular apoptosis and necrosis (Krogmann et al. 2016). We observed that pulmonary vascular endothelium of water buffalo did not show any TLR9 immunopositive reactivity which is similar to expression of TLR4 (Sethi et al. 2011) (Fig. 2.2). However, unlike TLR4, TLR9 expression was increased in the vascular endothelium during *P. multocida* induced lung inflammation in water buffalo (Sethi et al. 2011). The data suggest that increased TLR9 expression may handle the DNA released either from the phagocytosed bacteria or from the dying extracellular bacteria at the later stages of infection. However, the contradictory finding of TLR9 agonism in vascular biology remains the potential area of research. In horse and cattle, expression of TLR9 was clearly shown in vascular endothelial cells (Fig. 2.3), with some possible capillary staining, though if present it was much lighter in comparison (Schneberger et al. 2009, 2011).

Recently, TLR10 mRNA upregulation in THP-1 cell line has been linked with reactive oxygen species induced during hypoxia (Kim et al. 2010). TLR10 is expressed in lungs of human although at very low levels and immune cells viz neutrophils, macrophages, and dendritic cells (Chuang and Ulevitch 2001). We recently reported immuno-histochemical and immuno-electron microscopic data indicating expression of TLR10 in the pulmonary vascular endothelium of cattle along with other veterinary species (Balachandran et al. 2015). However, TLR10 immunohistological expression was much reduced in the *M. haemolytica* infected animals compared to the controls (Balachandran et al. 2015). The data suggest implications of TLR10 during *M. haemolytica* infection in terms of entry of bacteria to establish infection in cattle lung.

2.5 Pulmonary Intravascular Macrophages

One of the most striking differences between species in the lung capillaries is the presence of pulmonary intravascular macrophages (PIMs) attached to the microvascular endothelial cells of the lung. This population of macrophages colonizes the lung shortly after birth and are maintained through life in animals such as in horses, Artiodactyla, odontoceti, and cats, while in other species they may be induced under certain circumstances (reviewed in Schneberger et al. 2012). They are attached strongly to the endothelial cell layer through a number of darkly staining structures, which may complicate creating primary cell cultures. These attachments to the endothelial cells are unknown but are thought to be a glycosyl-phosphatidyl inositol anchor (Atwal et al. 1992; Singh et al. 1995; Singh and Atwal 1997). Attachment is seen on the thicker side of the endothelial cell, possibly to reduce interference with gas exchange, (Winkler 1988) and the cells range from 20 to 80 μM in size, with horse PIMs being the largest. While similar to other macrophages, PIMs possess a unique decoration of lipid/lipoprotein vesicles at or near the

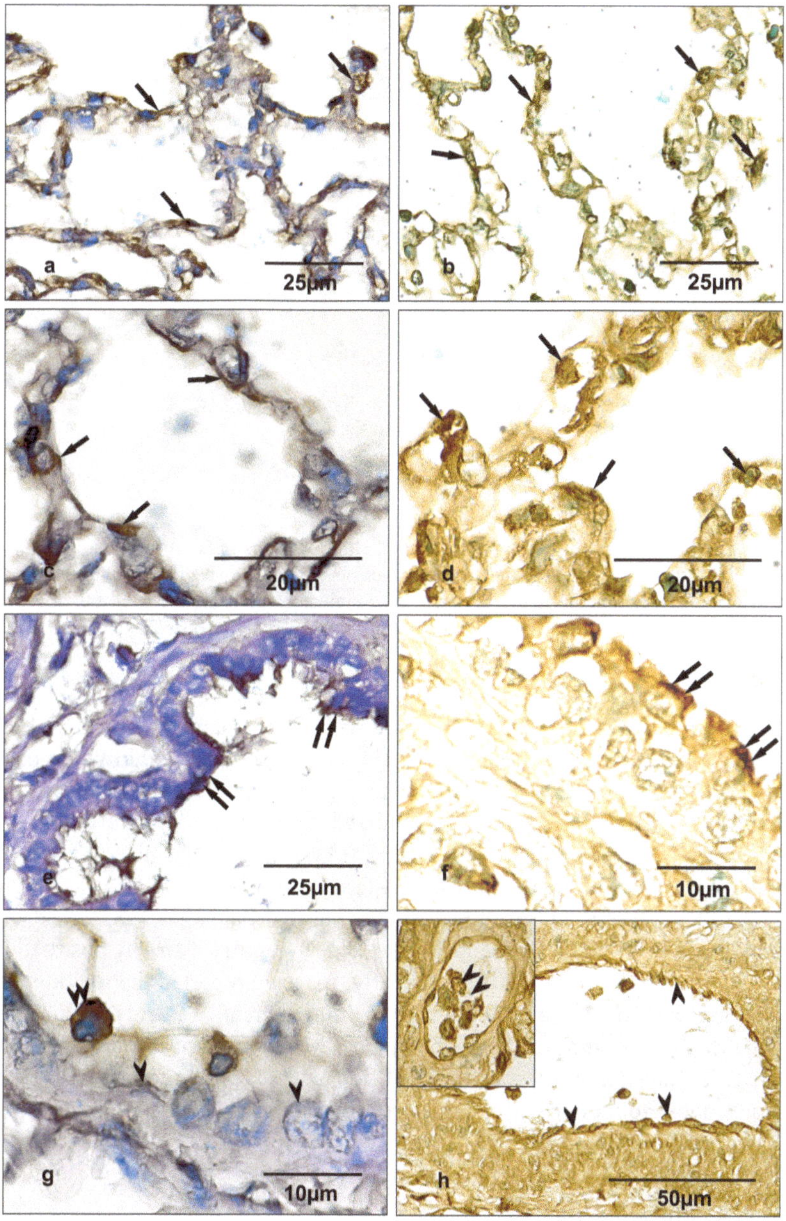

Fig. 2.2 (a) Normal lung section shows weak expression of TLR9 in alveolar septa (arrows) and (b) airway epithelium (double arrows), (c) but not in the endothelium of large blood vessels (arrowheads). (d) TLR9 expression was increased in the alveolar septa (arrows), (e) airway epithelium (double arrows) and (f) the endothelium of large blood vessels (arrowheads) in inflamed lungs. The blood cells (double arrowheads) in the control (c) and the infected lung (f) were also positive for TLR9 expression. IHC. Reprinted with permission from J Comp Pathol. 2011 Feb–Apr;144(2–3):135-44. doi: https://doi.org/10.1016/j.jcpa.2010.08.003

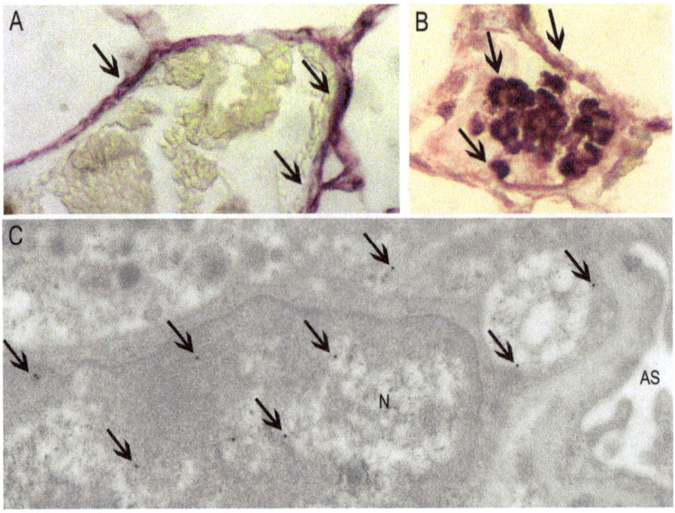

Fig. 2.3 TLR9 immunostaining of vascular endothelium in lungs of control (A) and LPS-treated horses (B) (noted by arrows). Immuno-electron microscopy (C) showed TLR9 staining in nucleus (N) of capillary endothelial cells in the lung (arrows). Original magnification A-B: ×400, C: ×10,000. Reprinted with permission from Anat Rec (Hoboken). 2009 Jul;292(7):1068–77. doi: https://doi.org/10.1002/ar.20927

cell surface membrane which are believed to help in phagocytosis (Atwal et al. 1992; Singh and Atwal 1997). Recently, PIMs were also identified in Water buffalo (Sethi et al. 2011; Fig. 2.4). Like other macrophage populations, PIMs are decorated with a number of TLR receptors (Wassef et al. 2004; Singh Suri et al. 2006; Schneberger et al. 2009, 2011; Sethi et al. 2011) (Fig. 2.5) and play a role in lung inflammation (Singh and de la Concha-Bermejillo 1998; Singh et al. 2004; Parbhakar et al. 2005), and phagocytosis (Atwal and Saldanha 1985). In disease, abrogation of PIMs can greatly reduce symptoms in recurrent airway obstruction in horses (Aharonson-Raz et al. 2012). Recently, we have reported the recruitment of PIMs in dogs that died due to acute necrotizing pancreatitis (Vrolyk et al. 2017). These data raise the intriguing possibility of PIM recruitment in domestic animal species that may not have constitutive PIMs and that the recruited PIMs may predispose these to higher susceptibility for lung inflammation and disease.

2.6 Culture of Endothelial Cells

Isolation and culture of equine endothelial cells has been described in detail (Lamar et al. 1986; MacEachern et al. 1997). Arterial endothelial cells are obtained by surgical removal of vessels and flushing them with PBS supplemented with antibiotics to remove any residual blood and potential bacterial contamination. Warmed DMEM with collagenase added is instilled in vessels which are clamped shut and

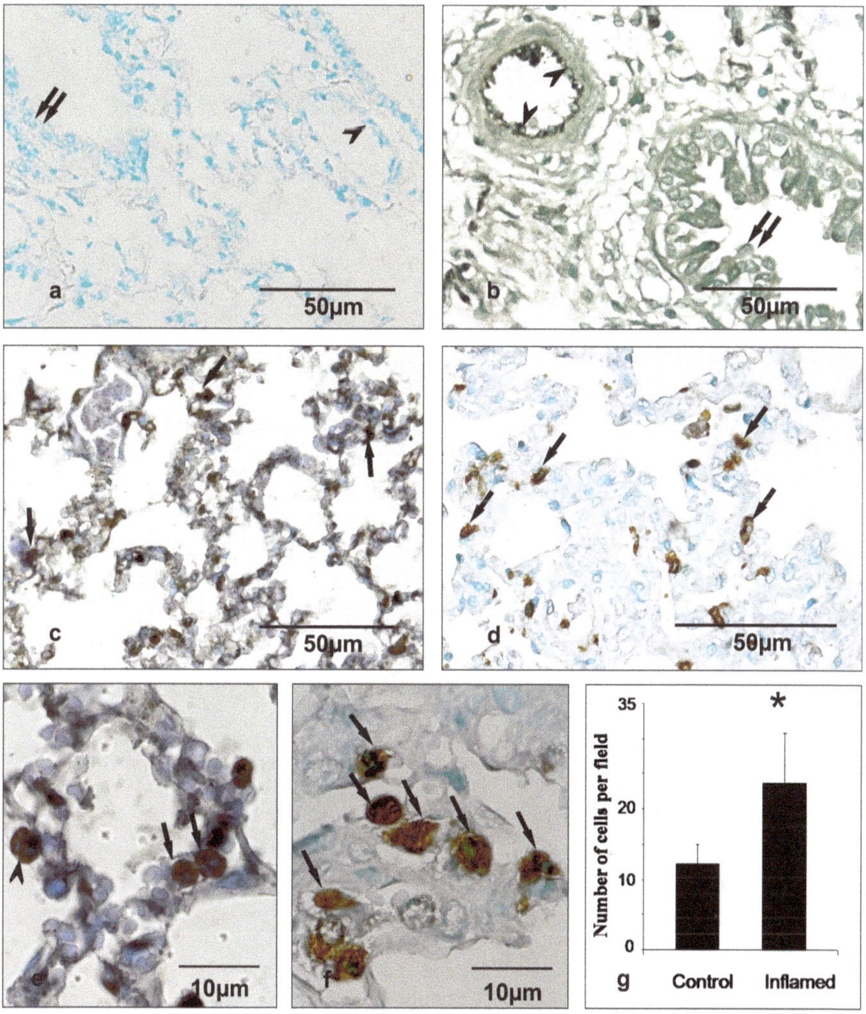

Fig. 2.4 (a) Lung section stained without primary antibody does not show any colour development in airways (double arrows) or bloodvessels (arrowheads). (b) Antibody specific for Factor VIII-related antigen labels vascular endothelium (arrowheads), but not the airways (double arrows). Macrophage antibody reacts with septal cells (arrows) in the normal lungs (c) and the infected lungs (d). (e) High magnification shows intravascular location of labelled macrophages (arrows) in the normal and (f) the infected lungs. An alveolar macrophage (arrowhead) is seen in (e). (g) The number of septal macrophages was significantly increased (P ¼ 0.001) in the inflamed lungs compared with the normal lungs. Reprinted with permission from J Comp Pathol. 2011 Feb–Apr;144(2–3):135–44. doi: https://doi.org/10.1016/j.jcpa.2010.08.003

incubated for 30–40 min at 37 °C. After incubation, vessels are rinsed emptied, media saved, and further washed with PBS and antibiotics. This wash is added to growth media and centrifuged to remove fluid, replacing it with final culture media of DMEM plus fetal calf serum and antibiotics. This is then cultured on 1% gelatin

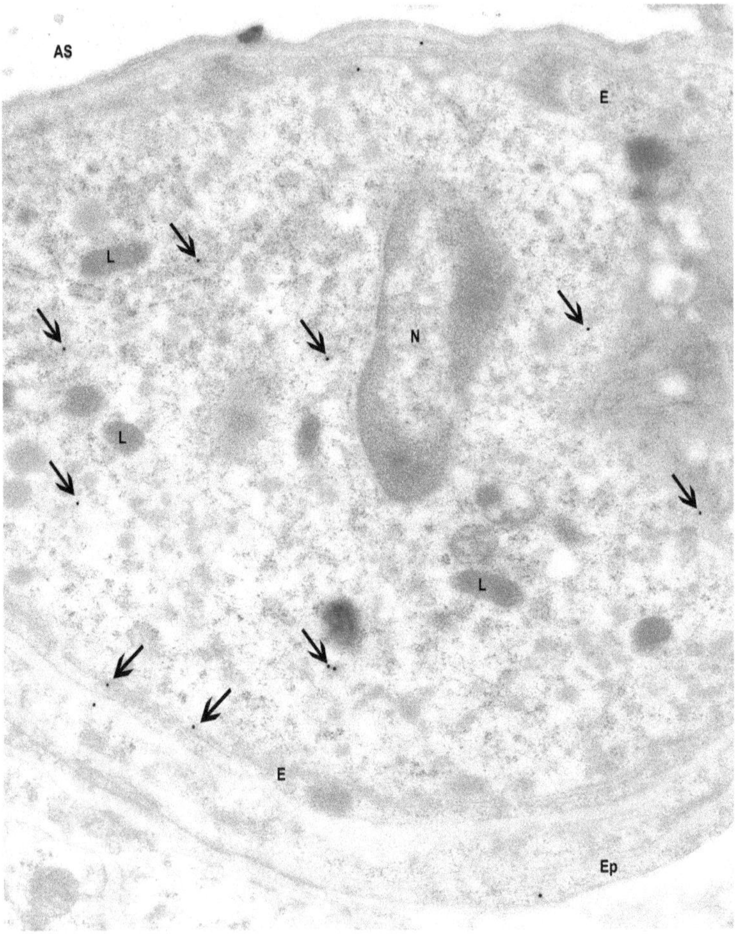

Fig. 2.5 TLR9 staining observed in a PIM and endothelium (E: arrows). L: Lysosomes; Ep: Epithelium; AS: Alveolar Space. Original magnification ×10,000. Anat Rec (Hoboken). 2009 Jul;292(7):1068–77. doi: https://doi.org/10.1002/ar.20927

coated plates, washing cells with PBS and replacing with fresh media until cells reach required density (MacEachern et al. 1997 for more details; Lamar et al. 1986).

This method yields primarily epithelial cells, but some smooth muscle cells will also be harvested in the process. Use of plasma-derived fetal bovine serum in media (10–20%) was able to inhibit growth of these smooth muscle cells, however, as opposed to equine-derived serum which stimulated this population as well as endothelial epithelium. It is believed that a lack of platelet-derived growth factors caused this inhibition of smooth muscle cells (Lamar et al. 1986). Cell type is verified by staining for vWF (Lamar et al. 1986).

Isolation procedures for lung microvascular endothelial cells will follow harvest and isolation techniques seen with capillaries in other tissues. Tissue is aseptically obtained and minced or otherwise mechanically disrupted in Hank's balanced salt solution, often with addition of antibiotics similar to arterial methods to ensure sterility. This suspension is then treated with collagenase to digest connective tissue and heated for 30 min at 37 °C before passing through a nylon mesh. Filtrate is combined with equal an equal volume of DMEM with 10% fetal bovine serum. After 10 min, cells are centrifuged at low speed, resuspended in DMEM+FBS, and passed through a 30 μM filter. The filter is washed for clumps of endothelial cells which are saved, while flow through is discarded (containing erythrocytes and stromal cells). Harvested cells are washed, re-centrifuged, re-suspended in HBSS and 10 ml gently layered onto 35 ml HBSS + 5% BSA in centrifuged and left stand at room temperature for 15 min. The bottom 20–25 ml containing enriched endothelial cells is saved while the upper layer containing stromal cells is discarded. Cells are centrifuged, resuspended in DMEM + 10% FBS plus antibiotics and endothelial cell growth supplement and plated into dishes coated with fibronectin (Bochsler et al. 1989).

While some microvascular cultures methods will suggest vWF as a method of detection of these cells, expression in microvascular endothelial cells is less robust, and there are no signs of WPb in these cells. Some other suggested markers include uptake of low density lipopeptide and angiotensin converting enzyme in bovine models (Chung-Welch et al. 1988). While not specifically tested in equine, CD34 has also been proven effective at labeling endothelial cells (Muller et al. 2002), and unlike vWF, it appears to be more strongly expressed in microvascular endothelial cells. Similarly, discrimination of different lung endothelial cells has been done using lectins that selectively bind to each endothelial population in several species including cattle (Schnitzer et al. 1994; Magee et al. 1994; Abdi et al. 1995), though these can vary between species, and to our knowledge no screening study of these has yet been done for the horse.

2.7 Conclusions

There are good amounts of data on the morphology and function of lung vascular endothelium in domestic animals. However, there are significant gaps in the molecular phenotyping of lung vascular endothelium in domestic animals. This will always remain the case because of the challenge of creating knock-outs or abilities to functionally block a molecule for precise functional phenotyping. The strength of the data obtained from domestic animal species is their closer physiological relevance to human's pulmonary physiology. There are, however, instances where the data on cells such as PIMs is opening new areas to investigate mechanisms of lung inflammation in rodent models and humans. Therefore, an increased focus on the comparative physiology of domestic animal species and less reliance on rodent models may provide more translational data to understand human lung vascular physiology. The increased use of methods such as electron microscopy especially the correlative light and electron microscopy will yield more precise localization of molecules of functional interest in lung capillary endothelium.

References

Abdi K, Kobzik L, Li X, Mentzer SJ (1995) Expression of membrane glycoconjugates on sheep lung endothelium. Lab Invest 72(4):445–452

Aderem A (2001) Role of Toll-like receptors in inflammatory response in macrophages. Crit Care Med 29(7 Suppl):S16–S18

Aharonson-Raz K, Lohmann KL, Townsend HG, Marques F, Singh B (2012) Pulmonary intravascular macrophages as proinflammatory cells in heaves, an asthma-like equine disease. Am J Physiol Lung Cell Mol Physiol 303(3):L189–L198. https://doi.org/10.1152/ajplung.00271.2011

Alvarez DF, King JA, Weber D, Addison E, Liedtke W, Townsley MI (2006) Transient receptor potential vanilloid 4-mediated disruption of the alveolar septal barrier: a novel mechanism of acute lung injury. Circ Res 99(9):988–995. https://doi.org/10.1161/01.RES.0000247065.11756.19

Andonegui G, Goyert SM, Kubes P (2002) Lipopolysaccharide-induced leukocyte-endothelial cell interactions: a role for CD14 versus toll-like receptor 4 within microvessels. J Immunol 169 (4):2111–2119

Andonegui G, Bonder CS, Green F, Mullaly SC, Zbytnuik L, Raharjo E, Kubes P (2003) Endothelium-derived Toll-like receptor-4 is the key molecule in LPS-induced neutrophil sequestration into lungs. J Clin Invest 111(7):1011–1020. https://doi.org/10.1172/JCI16510

Atwal OS, Saldanha KA (1985) Erythrophagocytosis in alveolar capillaries of goat lung: ultrastructural properties of blood monocytes. Acta Anat (Basel) 124(3-4):245–254

Atwal OS, Singh B, Staempfli H, Minhas K (1992) Presence of pulmonary intravascular macrophages in the equine lung: some structuro-functional properties. Anat Rec 234(4):530–540. https://doi.org/10.1002/ar.1092340408

Balachandran Y, Knaus S, Caldwell S, Singh B (2015) Toll-like receptor 10 expression in chicken, cattle, pig, dog, and rat lungs. Vet Immunol Immunopathol 168(3-4):184–192. https://doi.org/10.1016/j.vetimm.2015.09.007

Bauer S, Wagner H (2002) Bacterial CpG-DNA licenses TLR9. Curr Top Microbiol Immunol 270:145–154

Birch KA, Pober JS, Zavoico GB, Means AR, Ewenstein BM (1992) Calcium/calmodulin transduces thrombin-stimulated secretion: studies in intact and minimally permeabilized human umbilical vein endothelial cells. J Cell Biol 118(6):1501–1510

Birch KA, Ewenstein BM, Golan DE, Pober JS (1994) Prolonged peak elevations in cytoplasmic free calcium ions, derived from intracellular stores, correlate with the extent of thrombin-stimulated exocytosis in single human umbilical vein endothelial cells. J Cell Physiol 160 (3):545–554. https://doi.org/10.1002/jcp.1041600318

Birks EK, Mathieu-Costello O, Fu Z, Tyler WS, West JB (1994) Comparative aspects of the strength of pulmonary capillaries in rabbit, dog, and horse. Respir Physiol 97(2):235–246

Birks EK, Mathieu-Costello O, Fu Z, Tyler WS, West JB (1997) Very high pressures are required to cause stress failure of pulmonary capillaries in thoroughbred racehorses. J Appl Physiol (1985) 82(5):1584–1592

Bochsler PN, Slauson DO, Chandler SK, Suyemoto MM (1989) Isolation and characterization of equine microvascular endothelial cells in vitro. Am J Vet Res 50(10):1800–1805

Bonfanti R, Furie BC, Furie B, Wagner DD (1989) PADGEM (GMP140) is a component of Weibel-Palade bodies of human endothelial cells. Blood 73(5):1109–1112

Chuang T, Ulevitch RJ (2001) Identification of hTLR10: a novel human Toll-like receptor preferentially expressed in immune cells. Biochim Biophys Acta 1518(1-2):157–161

Chung-Welch N, Shepro D, Dunham B, Hechtman HB (1988) Prostacyclin and prostaglandin E2 secretions by bovine pulmonary microvessel endothelial cells are altered by changes in culture conditions. J Cell Physiol 135(2):224–234. https://doi.org/10.1002/jcp.1041350209

Cioffi DL, Moore TM, Schaack J, Creighton JR, Cooper DM, Stevens T (2002) Dominant regulation of interendothelial cell gap formation by calcium-inhibited type 6 adenylyl cyclase. J Cell Biol 157(7):1267–1278. https://doi.org/10.1083/jcb.200204022

Cioffi DL, Lowe K, Alvarez DF, Barry C, Stevens T (2009) TRPing on the lung endothelium: calcium channels that regulate barrier function. Antioxid Redox Signal 11(4):765–776. https://doi.org/10.1089/ARS.2008.2221

Costello ML, Mathieu-Costello O, West JB (1992) Stress failure of alveolar epithelial cells studied by scanning electron microscopy. Am Rev Respir Dis 145(6):1446–1455. https://doi.org/10.1164/ajrccm/145.6.1446

Cotran RS (1989) Endothelial cells. Textbook of rheumatology, 3rd edn. WB Saunders, Philadelphila, PA

Crapo JD, Barry BE, Gehr P, Bachofen M, Weibel ER (1982) Cell number and cell characteristics of the normal human lung. Am Rev Respir Dis 126(2):332–337. https://doi.org/10.1164/arrd.1982.126.2.332

Crapo JD, Young SL, Fram EK, Pinkerton KE, Barry BE, Crapo RO (1983) Morphometric characteristics of cells in the alveolar region of mammalian lungs. Am Rev Respir Dis 128 (2 Pt 2):S42–S46. https://doi.org/10.1164/arrd.1983.128.2P2.S42

Demiryurek AT, Wadsworth RM, Kane KA (1991) Effects of hypoxia on isolated intrapulmonary arteries from the sheep. Pulm Pharmacol 4(3):158–164

Demiryurek AT, Wadsworth RM, Kane KA, Peacock AJ (1993) The role of endothelium in hypoxic constriction of human pulmonary artery rings. Am Rev Respir Dis 147(2):283–290. https://doi.org/10.1164/ajrccm/147.2.283

Dornan JC, Meban C (1985) The capillary plexus in the gas exchange zone of human neonatal lung: an ultrastructural study. Thorax 40(10):787–792

Egan K, FitzGerald GA (2006) Eicosanoids and the vascular endothelium. Handb Exp Pharmacol 176 Pt 1:189–211

Elliott AR, Fu Z, Tsukimoto K, Prediletto R, Mathieu-Costello O, West JB (1992) Short-term reversibility of ultrastructural changes in pulmonary capillaries caused by stress failure. J Appl Physiol (1985) 73(3):1150–1158

Faure E, Equils O, Sieling PA, Thomas L, Zhang FX, Kirschning CJ, Polentarutti N, Muzio M, Arditi M (2000) Bacterial lipopolysaccharide activates NF-kappaB through toll-like receptor 4 (TLR-4) in cultured human dermal endothelial cells. Differential expression of TLR-4 and TLR-2 in endothelial cells. J Biol Chem 275(15):11058–11063

Fu Z, Costello ML, Tsukimoto K, Prediletto R, Elliott AR, Mathieu-Costello O, West JB (1992) High lung volume increases stress failure in pulmonary capillaries. J Appl Physiol (1985) 73 (1):123–133

Fuchs A, Weibel ER (1966) Morphometric study of the distribution of a specific cytoplasmatic organoid in the rat's endothelial cells. Zeitschrift fur Zellforschung und mikroskopische Anatomie (Vienna, Austria: 1948) 73(1):1–9

Furchgott RF, Vanhoutte PM (1989) Endothelium-derived relaxing and contracting factors. FASEB J 3(9):2007–2018

Furchgott RF, Zawadzki JV (1980) The obligatory role of endothelial cells in the relaxation of arterial smooth muscle by acetylcholine. Nature 288(5789):373–376

Gillespie JR, Tyler WS (1967) Quantitative electron microscopy of the interalveolar septa of the horse lung. Am Rev Respir Dis 95(3):477–483. https://doi.org/10.1164/arrd.1967.95.3.477

Heltianu C, Simionescu M, Simionescu N (1982) Histamine receptors of the microvascular endothelium revealed in situ with a histamine-ferritin conjugate: characteristic high-affinity binding sites in venules. J Cell Biol 93(2):357–364

Jian MY, King JA, Al-Mehdi AB, Liedtke W, Townsley MI (2008) High vascular pressure-induced lung injury requires P450 epoxygenase-dependent activation of TRPV4. Am J Respir Cell Mol Biol 38(4):386–392. https://doi.org/10.1165/rcmb.2007-0192OC

Kaiser L, Sparks HV Jr (1986) Mediation of flow-dependent arterial dilation by endothelial cells. Circ Shock 18(2):109–114

Kelly JJ, Moore TM, Babal P, Diwan AH, Stevens T, Thompson WJ (1998) Pulmonary microvascular and macrovascular endothelial cells: differential regulation of Ca2+ and permeability. Am J Physiol 274(5 Pt 1):L810–L819

Kim D, Kim YJ, Koh HS, Jang TY, Park HE, Kim JY (2010) Reactive oxygen species enhance TLR10 expression in the human monocytic cell line THP-1. Int J Mol Sci 11(10):3769–3782. https://doi.org/10.3390/ijms11103769

Kindig CA, Erickson HH, Poole DC (2000) Dissociation of exercise-induced pulmonary hemorrhage and pulmonary artery pressure via nitric oxide synthase inhibition. J Equine Vet Sci 20 (11):715–721. https://doi.org/10.1016/S0737-0806(00)80182-3

King J, Hamil T, Creighton J, Wu S, Bhat P, McDonald F, Stevens T (2004) Structural and functional characteristics of lung macro- and microvascular endothelial cell phenotypes. Microvasc Res 67(2):139–151. https://doi.org/10.1016/j.mvr.2003.11.006

Koller A, Sun D, Huang A, Kaley G (1994) Corelease of nitric oxide and prostaglandins mediates flow-dependent dilation of rat gracilis muscle arterioles. Am J Physiol 267(1 Pt 2):H326–H332

Krogmann AO, Lusebrink E, Steinmetz M, Asdonk T, Lahrmann C, Lutjohann D, Nickenig G, Zimmer S (2016) Proinflammatory stimulation of Toll-like receptor 9 with high dose CpG ODN 1826 impairs endothelial regeneration and promotes atherosclerosis in mice. PLoS One 11(1):e0146326. https://doi.org/10.1371/journal.pone.0146326

Kuebler WM, Ying X, Singh B, Issekutz AC, Bhattacharya J (1999) Pressure is proinflammatory in lung venular capillaries. J Clin Invest 104(4):495–502. https://doi.org/10.1172/JCI6872

Lamar CH, Turek JJ, Bottoms GD, Fessler JF (1986) Equine endothelial cells in vitro. Am J Vet Res 47(4):956–958

Lorant DE, Patel KD, McIntyre TM, McEver RP, Prescott SM, Zimmerman GA (1991) Coexpression of GMP-140 and PAF by endothelium stimulated by histamine or thrombin: a juxtacrine system for adhesion and activation of neutrophils. J Cell Biol 115(1):223–234

Lowenstein CJ, Morrell CN, Yamakuchi M (2005) Regulation of Weibel-Palade body exocytosis. Trends Cardiovasc Med 15(8):302–308. https://doi.org/10.1016/j.tcm.2005.09.005

Ludwig MG, Seuwen K (2002) Characterization of the human adenylyl cyclase gene family: cDNA, gene structure, and tissue distribution of the nine isoforms. J Recept Signal Transduct Res 22(1-4):79–110. https://doi.org/10.1081/RRS-120014589

Lundberg JM, Martling CR, Hökfelt T (1988) Airways, oral cavity and salivary glands: classical transmitters and peptides in sensory and autonomic motor neurons. In: Björklund A, Hökfelt T, Owman C (eds) Handbook of chemical neuroanatomy, vol 6. Elsevier Science, New York, NY, pp 391–444

Lymboussaki A, Partanen TA, Olofsson B, Thomas-Crusells J, Fletcher CD, de Waal RM, Kaipainen A, Alitalo K (1998) Expression of the vascular endothelial growth factor C receptor VEGFR-3 in lymphatic endothelium of the skin and in vascular tumors. Am J Pathol 153 (2):395–403. https://doi.org/10.1016/S0002-9440(10)65583-6

MacEachern KE, Smith GL, Nolan AM (1997) Methods for the isolation, culture and characterisation of equine pulmonary artery endothelial cells. Res Vet Sci 62(2):147–152

MacEachern KE, Smith GL, Nolan AM (2004) Characteristics of the in vitro hypoxic pulmonary vasoconstrictor response in isolated equine and bovine pulmonary arterial rings. Vet Anaesth Analg 31(4):239–249. https://doi.org/10.1111/j.1467-2995.2004.00176.x

Magee JC, Stone AE, Oldham KT, Guice KS (1994) Isolation, culture, and characterization of rat lung microvascular endothelial cells. Am J Physiol 267(4 Pt 1):L433–L441

Mantovani A, Bussolino F, Dejana E (1992) Cytokine regulation of endothelial cell function. FASEB J 6(8):2591–2599

Marchesi VT (1961) The site of leucocyte emigration during inflammation. Q J Exp Physiol Cogn Med Sci 46:115–118

Michel RP, Hu F, Meyrick BO (1995) Myoendothelial junctional complexes in postobstructive pulmonary vasculopathy: a quantitative electron microscopic study. Exp Lung Res 21 (3):437–452

Millar FR, Summers C, Griffiths MJ, Toshner MR, Proudfoot AG (2016) The pulmonary endothelium in acute respiratory distress syndrome: insights and therapeutic opportunities. Thorax 71 (5):462–473. https://doi.org/10.1136/thoraxjnl-2015-207461

Minami T, Sugiyama A, Wu SQ, Abid R, Kodama T, Aird WC (2004) Thrombin and phenotypic modulation of the endothelium. Arterioscler Thromb Vasc Biol 24(1):41–53. https://doi.org/10.1161/01.ATV.0000099880.09014.7D

Muller AM, Nesslinger M, Skipka G, Muller KM (2002) Expression of CD34 in pulmonary endothelial cells in vivo. Pathobiology 70(1):11–17

Nishi Y, Kitamura N, Otani M, Hondo E, Taguchi K, Yamada J (2000) Distribution of capsaicin-sensitive substance P- and calcitonin gene-related peptide-immunoreactive nerves in bovine respiratory tract. Ann Anat 182(4):319–326. https://doi.org/10.1016/S0940-9602(00)80004-7

Ochoa CD, Wu S, Stevens T (2010) New developments in lung endothelial heterogeneity: Von Willebrand factor, P-selectin, and the Weibel-Palade body. Semin Thromb Hemost 36 (3):301–308. https://doi.org/10.1055/s-0030-1253452

Ofori-Acquah SF, King J, Voelkel N, Schaphorst KL, Stevens T (2008) Heterogeneity of barrier function in the lung reflects diversity in endothelial cell junctions. Microvasc Res 75 (3):391–402. https://doi.org/10.1016/j.mvr.2007.10.006

Pan J, Xia L, McEver RP (1998) Comparison of promoters for the murine and human P-selectin genes suggests species-specific and conserved mechanisms for transcriptional regulation in endothelial cells. J Biol Chem 273(16):10058–10067

Parbhakar OP, Duke T, Townsend HG, Singh B (2005) Depletion of pulmonary intravascular macrophages partially inhibits lipopolysaccharide-induced lung inflammation in horses. Vet Res 36(4):557–569. https://doi.org/10.1051/vetres:2005016

Parker JC, Yoshikawa S (2002) Vascular segmental permeabilities at high peak inflation pressure in isolated rat lungs. Am J Physiol Lung Cell Mol Physiol 283(6):L1203–L1209. https://doi.org/10.1152/ajplung.00488.2001

Parker JC, Stevens T, Randall J, Weber DS, King JA (2006) Hydraulic conductance of pulmonary microvascular and macrovascular endothelial cell monolayers. Am J Physiol Lung Cell Mol Physiol 291(1):L30–L37. https://doi.org/10.1152/ajplung.00317.2005

Parthasarathi K, Ichimura H, Monma E, Lindert J, Quadri S, Issekutz A, Bhattacharya J (2006) Connexin 43 mediates spread of Ca2+-dependent proinflammatory responses in lung capillaries. J Clin Invest 116(8):2193–2200. https://doi.org/10.1172/JCI26605

Pascoe JR, Ferraro GL, Cannon JH, Arthur RM, Wheat JD (1981) Exercise-induced pulmonary hemorrhage in racing thoroughbreds: a preliminary study. Am J Vet Res 42(5):703–707

Pelletier N, Leith DE (1993) Cardiac output but not high pulmonary artery pressure varies with FIO2 in exercising horses. Respir Physiol 91(1):83–97

Pober JS, Cotran RS (1990) Cytokines and endothelial cell biology. Physiol Rev 70(2):427–451

Pober JS, Sessa WC (2007) Evolving functions of endothelial cells in inflammation. Nat Rev Immunol 7(10):803–815. https://doi.org/10.1038/nri2171

Prasain N, Stevens T (2009) The actin cytoskeleton in endothelial cell phenotypes. Microvasc Res 77(1):53–63. https://doi.org/10.1016/j.mvr.2008.09.012

Prescott SM, Zimmerman GA, McIntyre TM (1984) Human endothelial cells in culture produce platelet-activating factor (1-alkyl-2-acetyl-sn-glycero-3-phosphocholine) when stimulated with thrombin. Proc Natl Acad Sci U S A 81(11):3534–3538

Rees DD, Palmer RM, Moncada S (1989) Role of endothelium-derived nitric oxide in the regulation of blood pressure. Proc Natl Acad Sci U S A 86(9):3375–3378

Sayner S, Stevens T (2006) Soluble adenylate cyclase reveals the significance of compartmentalized cAMP on endothelial cell barrier function. Biochem Soc Trans 34(Pt 4):492–494. https://doi.org/10.1042/BST0340492

Sayner SL, Alexeyev M, Dessauer CW, Stevens T (2006) Soluble adenylyl cyclase reveals the significance of cAMP compartmentation on pulmonary microvascular endothelial cell barrier. Circ Res 98(5):675–681. https://doi.org/10.1161/01.RES.0000209516.84815.3e

Schenkel AR, Mamdouh Z, Chen X, Liebman RM, Muller WA (2002) CD99 plays a major role in the migration of monocytes through endothelial junctions. Nat Immunol 3(2):143–150. https://doi.org/10.1038/ni749

Schneberger D, Caldwell S, Suri SS, Singh B (2009) Expression of toll-like receptor 9 in horse lungs. Anat Rec (Hoboken) 292(7):1068–1077. https://doi.org/10.1002/ar.20927

Schneberger D, Lewis D, Caldwell S, Singh B (2011) Expression of toll-like receptor 9 in lungs of pigs, dogs and cattle. Int J Exp Pathol 92(1):1–7. https://doi.org/10.1111/j.1365-2613.2010.00742.x

Schneberger D, Aharonson-Raz K, Singh B (2012) Pulmonary intravascular macrophages and lung health: what are we missing? Am J Physiol Lung Cell Mol Physiol 302(6):L498–L503. https://doi.org/10.1152/ajplung.00322.2011

Schneeberger EE (1982) Structure of intercellular junctions in different segments of the intrapulmonary vasculature. Ann N Y Acad Sci 384:54–63

Schnitzer JE, Siflinger-Birnboim A, Del Vecchio PJ, Malik AB (1994) Segmental differentiation of permeability, protein glycosylation, and morphology of cultured bovine lung vascular endothelium. Biochem Biophys Res Commun 199(1):11–19

Sessa WC (2004) eNOS at a glance. J Cell Sci 117(Pt 12):2427–2429. https://doi.org/10.1242/jcs.01165

Sethi RS, Brar RS, Singh O, Singh B (2011) Immunolocalization of pulmonary intravascular macrophages, TLR4, TLR9 and IL-8 in normal and Pasteurella multocida-infected lungs of water buffalo (Bubalus bubalis). J Comp Pathol 144(2–3):135–144. https://doi.org/10.1016/j.jcpa.2010.08.003

Singh B, Atwal OS (1997) Ultrastructural and immunocytochemical study of the pulmonary intravascular macrophages of Escherichia coli lipopolysaccharide-treated sheep. Anat Rec 247(2):214–224

Singh B, de la Concha-Bermejillo A (1998) Gadolinium chloride removes pulmonary intravascular macrophages and curtails the degree of ovine lentivirus-induced lymphoid interstitial pneumonia. Int J Exp Pathol 79(3):151–162

Singh B, Ireland WP, Minhas K, Atwal OS (1995) Surface coat of sheep pulmonary intravascular macrophages: reconstitution, and implication of a glycosyl-phosphatidylinositol anchor. Anat Rec 243(4):466–478. https://doi.org/10.1002/ar.1092430409

Singh B, Pearce JW, Gamage LN, Janardhan K, Caldwell S (2004) Depletion of pulmonary intravascular macrophages inhibits acute lung inflammation. Am J Physiol Lung Cell Mol Physiol 286(2):L363–L372. https://doi.org/10.1152/ajplung.00003.2003

Singh Suri S, Janardhan KS, Parbhakar O, Caldwell S, Appleyard G, Singh B (2006) Expression of toll-like receptor 4 and 2 in horse lungs. Vet Res 37(4):541–551. https://doi.org/10.1051/vetres:2006017

Sirianni FE, Chu FS, Walker DC (2003) Human alveolar wall fibroblasts directly link epithelial type 2 cells to capillary endothelium. Am J Respir Crit Care Med 168(12):1532–1537. https://doi.org/10.1164/rccm.200303-371OC

Stevens T, Nakahashi Y, Cornfield DN, McMurtry IF, Cooper DM, Rodman DM (1995) Ca(2+)-inhibitable adenylyl cyclase modulates pulmonary artery endothelial cell cAMP content and barrier function. Proc Natl Acad Sci U S A 92(7):2696–2700

Takeda K, Kaisho T, Akira S (2003) Toll-like receptors. Annu Rev Immunol 21:335–376. https://doi.org/10.1146/annurev.immunol.21.120601.141126

Townsley MI (2012) Structure and composition of pulmonary arteries, capillaries, and veins. Compr Physiol 2(1):675–709. https://doi.org/10.1002/cphy.c100081

Tsukimoto K, Mathieu-Costello O, Prediletto R, Elliott AR, West JB (1991) Ultrastructural appearances of pulmonary capillaries at high transmural pressures. J Appl Physiol (1985) 71 (2):573–582

Turek JJ, Lamar CH, Fessler JF, Bottoms GD (1987) Ultrastructure of equine endothelial cells exposed to endotoxin and flunixin meglumine and equine neutrophils. Am J Vet Res 48 (9):1363–1366

Vrolyk V, Wobeser BK, Al-Dissi AN, Carr A, Singh B (2017) Lung inflammation associated with clinical acute necrotizing pancreatitis in dogs. Vet Pathol 54(1):129–140. https://doi.org/10.1177/0300985816646432

Walker DC, MacKenzie A, Hosford S (1994) The structure of the tricellular region of endothelial tight junctions of pulmonary capillaries analyzed by freeze-fracture. Microvasc Res 48 (3):259–281. https://doi.org/10.1006/mvre.1994.1054

Walker DC, Behzad AR, Chu F (1995) Neutrophil migration through preexisting holes in the basal laminae of alveolar capillaries and epithelium during streptococcal pneumonia. Microvasc Res 50(3):397–416. https://doi.org/10.1006/mvre.1995.1067

Wang JW, Eikenboom J (2010) Von Willebrand disease and Weibel-Palade bodies. Hamostaseologie 30(3):150–155

Wassef A, Janardhan K, Pearce JW, Singh B (2004) Toll-like receptor 4 in normal and inflamed lungs and other organs of pig, dog and cattle. Histol Histopathol 19(4):1201–1208

West JB (2000) Invited review: pulmonary capillary stress failure. J Appl Physiol (1985) 89 (6):2483–2489. discussion 2497

West JB, Mathieu-Costello O, Jones JH, Birks EK, Logemann RB, Pascoe JR, Tyler WS (1993) Stress failure of pulmonary capillaries in racehorses with exercise-induced pulmonary hemorrhage. J Appl Physiol (1985) 75(3):1097–1109

Winkler GC (1988) Pulmonary intravascular macrophages in domestic animal species: review of structural and functional properties. Am J Anat 181(3):217–234. https://doi.org/10.1002/aja.1001810302

Wu S, Haynes J Jr, Taylor JT, Obiako BO, Stubbs JR, Li M, Stevens T (2003) Cav3.1 (alpha1G) T-type Ca2+ channels mediate vaso-occlusion of sickled erythrocytes in lung microcirculation. Circ Res 93(4):346–353. https://doi.org/10.1161/01.RES.0000087148.75363.8F

Wu S, Jian MY, Xu YC, Zhou C, Al-Mehdi AB, Liedtke W, Shin HS, Townsley MI (2009) Ca2+ entry via alpha1G and TRPV4 channels differentially regulates surface expression of P-selectin and barrier integrity in pulmonary capillary endothelium. Am J Physiol Lung Cell Mol Physiol 297(4):L650–L657. https://doi.org/10.1152/ajplung.00015.2009

Xu SF, Wang P, Liang ZX, Sun JP, Zhao XW, Li AM, Chen LA (2011) Investigation of inflammatory responses of pulmonary microvascular endothelial cells induced by lipopolysaccharide and mechanism. Zhonghua Jie He He Hu Xi Za Zhi 34(11):816–820

Yin J, Hoffmann J, Kaestle SM, Neye N, Wang L, Baeurle J, Liedtke W, Wu S, Kuppe H, Pries AR, Kuebler WM (2008) Negative-feedback loop attenuates hydrostatic lung edema via a cGMP-dependent regulation of transient receptor potential vanilloid 4. Circ Res 102 (8):966–974. https://doi.org/10.1161/CIRCRESAHA.107.168724

Chapter 3
The Pulmonary Vascular Barrier: Insights into Structure, Function, and Regulatory Mechanisms

Kaushik Parthasarathi

Abstract Pulmonary blood vessels act as a well-regulated barrier to the flux of fluid and solutes between the lumen and the air space. Perturbation of the barrier function results in excessive fluid leak into the interstitium and alveoli, and impairs gas exchange. Recent studies provide deeper insight into the precise control mechanisms involved in the regulation of the barrier. This chapter will highlight these mechanisms and discuss the current understanding on the fluid and solute transport pathways across the vascular endothelial layer. In addition, the chapter summarizes the contributions of extra-endothelial structures such as pericytes and glycocalyx in regulating fluid flux across pulmonary vessels. The chapter concludes with an analysis on the impact of pulmonary endothelial heterogeneity and experimental models on current interpretations of barrier function and regulatory mechanisms.

3.1 Introduction

The pulmonary vasculature forms a semi-permeable barrier and acts as the first line of defense that limits the movement of fluid and macromolecules from blood into the interstitium. When this protective barrier is impaired, fluid leaks across the vascular endothelium, resulting in accumulation of the fluid in the interstitial space and subsequently entering the pulmonary air spaces. Excessive leak that overwhelms the fluid clearance mechanisms of the pulmonary epithelial barrier (Vadasz and Sznajder 2017) leads to pulmonary edema.

Fluid flux across the pulmonary vascular barrier and development of pulmonary edema could be of either cardiogenic or non-cardiogenic origin. Cardiogenic pulmonary edema results from impairment of cardiac function, including myocardial

K. Parthasarathi (✉)
Department of Physiology, University of Tennessee Health Science Center, Memphis, TN, USA
e-mail: kparthas@uthsc.edu

K. Parthasarathi (ed.), *Molecular and Functional Insights Into the Pulmonary Vasculature*, Advances in Anatomy, Embryology and Cell Biology 228,
https://doi.org/10.1007/978-3-319-68483-3_3

infarction and congestive heart failure, which lead to elevation of hydrostatic pressure gradients in pulmonary microvessels (Cosentini et al. 2009). The increased pressure differences result in the leakage of fluid with low protein content, typically less than 65% compared to that in plasma into the interstitial space (Ware et al. 2010). In contrast, non-cardiogenic edema results from alterations in the permeability properties of the pulmonary vascular barrier. In response, fluid with protein content similar to that in plasma leaks into the interstitium (Ware et al. 2010; Weidenfeld and Kuebler 2017). Excessive accumulation of proteinaceous fluid in air spaces of the lung is a characteristic feature of acute lung injury and its more severe form, acute respiratory distress syndrome (Gonzales et al. 2015; Matthay et al. 2017; Wang et al. 2017). The accumulating fluid impairs gas exchange, resulting in arterial hypoxemia and hypercapnia (Matthay et al. 2012). The ensuing tissue hypoxia can in itself induce inflammation (Eltzschig and Carmeliet 2011; Eltzschig et al. 2014) and result in organ dysfunction. Given the important role of the pulmonary vascular barrier in limiting fluid flux, maintenance and tight regulation of the barrier is critical to maintaining proper lung function and homeostasis.

This chapter provides a broad outline of the current understanding of the features of the pulmonary vascular barrier, mechanisms that result in failure of the interendothelial junctions, and the role of the actin cytoskeleton. Mechanisms leading to the lung vascular barrier breakdown and emerging therapeutic mechanisms aimed at reversing excessive fluid flux are briefly discussed. Several recent reviews provide additional in-depth information on specific topics and are referred to throughout the article.

3.2 Transcellular Fluid–Flux Pathway

3.2.1 Aquaporins

Fluid and solute flux across the pulmonary vasculature could utilize either the paracellular or transcellular pathway. The transcellular fluid flux is considered to contribute minimally to the transport of water across the endothelial barrier. Transport via this pathway is mediated either via aquaporins or caveolae. Lung endothelial plasma membranes express aquaporin-1 (Schnitzer and Oh 1996; Bai et al. 1999; King et al. 2002). However, the role of aquaporin-1 as a mediator of osmotically driven water transport across the pulmonary vascular endothelium has remained controversial, due to the limited loss of fluid transport functions in aquaporin null mice (Verkman 2007). However, newer evidence point to a modification in the expression of aquaporin-1 in human lungs with diffuse alveolar damage (Pires-Neto et al. 2016). Whether the newer studies support or discount the role of aquaporin-1 in lung fluid balance remain to be seen.

3.2.2 Caveolae

In contrast to aquaporins, there is a growing body of evidence that suggest a role for caveolae in lung fluid balance. Caveolae are plasma membrane-associated flask-shaped pits, averaging 55 nm in diameter (Richter et al. 2008). Caveolae are abundant in multiple cell types including endothelial cells and are involved in diverse functions including mechanosensing, lipid regulation, and endocytosis (Heimerl et al. 2008; Hansen and Nichols 2010; Sinha et al. 2011; Parton and del Pozo 2013). Loss of caveolae results in cardiovascular dysfunction due to impaired nitric oxide and calcium signaling (Drab et al. 2001). In lungs, absence of proteins critical to the formation of caveolae causes dysregulation of endothelial proliferation and fibrosis (Drab et al. 2001; Hansen et al. 2013).

Endothelial caveolae play an important role in albumin transcytosis via endocytosis, trafficking across the cell, and exocytosis at the basolateral plasma membrane. Signaling associated with caveolae-mediated albumin transcytosis is increasingly better understood since the seminal descriptions of the process by Palade (Palade 1953, 1961; Milici et al. 1987). Caveolae coat proteins, caveolin-1, caveolin-2, and caveolin-3, mediate caveolae-dependent signaling (Le Lay and Kurzchalia 2005). Recent reviews provide extensive details on the structure and function of caveolins (Maniatis et al. 2012; Sowa 2012; Hansen et al. 2013; Parton and del Pozo 2013). Lack of caveolin-1 results in loss of caveolae in endothelial cells, which leads to reduced transcytosis of albumin (Li et al. 2013). Interestingly, loss of caveolin-1 results in elevated permeability via the paracellular route (Schubert et al. 2002), indicating that the two pathways regulating endothelial permeability are likely intertwined.

Caveolae-mediated albumin transcytosis is mediated by interaction of caveolin-1 with GTP-ases, rac, and dynamin (Armstrong et al. 2012) and requires integrated actions by actin cytoskeleton-associated proteins, including filamin A (Sverdlov et al. 2009). The transcytosis rate in pulmonary microvascular endothelial cells can be modulated by inflammatory agents such as lipopolysaccharide, thrombin, high-mobility group box protein 1, and paraquat, which act through intermediates such as nuclear factor-κB, acid spingomyelinase, advanced glycation end products, and C-src, respectively (Tiruppathi et al. 2008; Wang et al. 2015; Kuebler et al. 2016; Shang et al. 2016; Huang and He 2017). The findings from these reports suggest that albumin transcytosis plays a key role in augmenting transvascular fluid flux under inflammatory conditions. As discussed in other sections of this chapter, paracellular pathways for fluid flux are also enhanced by proinflammatory mediators. While arguments of the dominance of one mechanism over another have been posited, it is very likely that both mechanisms may act in tandem toward regulating fluid flux across the endothelial barrier. Thus, resolution of inflammation-induced increase in fluid leak may require a holistic approach targeting the elevated fluid flux via both para- and trans-cellular routes.

3.3 Paracellular Fluid–Flux Pathway

3.3.1 Adherens Junctions

Fluid flux via the paracellular route across the pulmonary vascular barrier is regulated by a complex set of interacting proteins. The interendothelial junction is populated by junctional complexes that include adherens junctions, tight junctions, and gap junctions. Adherens junctions of the vascular endothelium are primarily composed of vascular endothelial (VE)-cadherin, the extracellular domain of which binds homotypically at the interendothelial junctions (Taveau et al. 2008). Disruption of this VE-cadherin-mediated cell–cell interaction results in augmented permeability across the microvessel barrier (Corada et al. 2002), suggesting the critical role of VE-cadherin in the maintenance of this barrier. On the cytoplasmic side, VE-cadherin forms a complex with several catenin subtypes including α, β, γ, and p120 catenin. The catenins are targets for cell signaling during inflammation and thus serve as regulatory switches in the modulation of microvessel permeability. Recent reviews provide extensive details on the structure of VE-cadherin and its interaction with catenins (Quadri 2012; Vogel and Malik 2012; Dejana and Orsenigo 2013; Gavard 2014). Coupling between the VE-cadherin–catenin complex and actin cytoskeletal network lends stability to the adherens junction complex (Dorland and Huveneers 2017). Dynamic interaction between the actin cytoskeleton and the VE-cadherin–catenin complex (Oldenburg and de Rooij 2014; Schnittler et al. 2014) facilitates both the physiological and pathological modulation of the interendothelial junction strength.

Given the complex interplay among various components of the adherens junctions, it is no surprise that perturbations of any one component elicited via changes in protein expression can lead to disassembly of the junctions, and, thus, disruption of the pulmonary vascular barrier (Gao et al. 2000; Dejana and Orsenigo 2013). In addition to expression changes, redistribution of junctional components from the plasma membrane sites could also induce similar disruptions (Sawant et al. 2011). In this regard, redistribution of VE-cadherin from the membrane and, thus, loss of VE-cadherin homophilic interactions could lead to loss of barrier integrity. Relevant mechanisms that initiate the redistribution have been studied extensively (Iyer et al. 2004; Fainaru et al. 2008; Herwig et al. 2008; Ieguchi et al. 2013). Modulations in cytosolic calcium levels play an important role in inducing changes in VE-cadherin homophilic interaction and its redistribution. Calcium is essential for the maintenance of VE-cadherin homophilic interactions, with low cytosolic calcium levels leading to disassembly of VE-cadherins (Gao et al. 2000). However, calcium plays the role of a double-edged sword as an increase in endothelial cytosolic Ca^{2+} also leads to destabilization of VE-cadherin interactions.

In this context, mechanisms for cytosolic calcium increase include store-operated calcium entry, which is activated in response to depletion of endoplasmic calcium stores. Blocking store-operated calcium entry by inhibiting the proteins that mediate this process, including STIM1, the endoplasmic store calcium sensor, and Orai1, the plasma membrane pore-forming component, (Ambudkar et al. 2017), both attenuate fluid leak across the pulmonary microvascular endothelium (Wang et al. 2016; Yazbeck et al. 2017). Interestingly, modulating events both upstream and downstream of STIM1-Orai1 also bears upon endothelial barrier strength. Inositol 1,4,5-trisphosphate, the messenger involved in triggering calcium release from endoplasmic reticulum stores, which is an early event leading to store-operated Ca^{2+} entry (Wu et al. 2014), causes clustering of its receptors on the endoplasmic reticulum. Inhibiting the receptor clustering blocks endoplasmic reticulum calcium release and blunts endothelial permeability (Geyer et al. 2015). Transient receptor potential channels facilitate calcium entry and a subset of these are triggered by store-operated calcium entry. One such channel that is thought to be downstream of STIM1 and Orai1 is transient receptor potential canonical 1 (Ambudkar et al. 2017), whose function is relevant in the context of pulmonary microvessels (Sundivakkam et al. 2012). Activation of this channel and the resulting increase in cytosolic calcium leads to perturbation of the microvascular barrier (Ahmmed and Malik 2005). However, the exact contributions of transient receptor potential channels to microvascular permeability and the specific mechanisms remain to be teased out further (Villalta and Townsley 2013; Malczyk et al. 2017). The presence of multiple transient receptor potential channels in the pulmonary vasculature, the heterogeneity in channel expression among the different regions of the pulmonary vasculature, and the overlap of triggering events complicate the delineation of channel contribution to pulmonary microvessel barrier disruption.

Increase in endothelial cytosolic calcium activate different signaling pathways that impact on the integrity of the adherens junctions. For an extensive delineation and discussion of the pathways, the reader is referred to comprehensive reviews on the topic (Tiruppathi et al. 2006; Ambudkar et al. 2017). Recent studies show that the integrity of adherens junctions is modulated precisely via tyrosine phosphorylation of the various components of these junctions, through the action of kinases, which in themselves are activated by elevated levels of endothelial cytosolic calcium (Adam 2015). For example, increases in cytosolic calcium by thrombin activate protein kinase C subtypes, which phosphorylate p120 catenin (Lucas et al. 2012; Vandenbroucke St Amant et al. 2012; Bijli et al. 2016). In addition to p120 catenin, phosphorylation of β-catenin and indeed VE-cadherin itself has also been shown to result in disassembly of adherens junctions (Cai et al. 2017; Muramatsu et al. 2017; Soni et al. 2017). Interestingly, tyrosine phosphorylation of VE-cadherin and catenins is only one of the multiple mechanisms associated with adherens junction disruption. Internalization of VE-cadherin and enzymatic cleavage of VE-cadherin also contribute to disassembly of adherens junctions and are reviewed elsewhere (Dejana et al. 2008). Intriguingly, it has recently been shown that phosphorylation of VE-cadherin in itself is only a necessary, but not a sufficient condition to induce disruption of adherens junctions (Orsenigo et al. 2012). In addition to phosphorylation, it has been posited that the presence of inflammatory agents is also required in order for an increase in microvessel permeability to occur. These reports suggest that our understanding of the barrier regulation via modulating adherens junction proteins is still incomplete and requires further elucidation.

3.3.2 Tight Junctions

Intertwined among the VE-cadherin interendothelial bridges are tight junction proteins, including claudins, occludins, and junction adhesion molecules (Tsukita et al. 2001; Dejana 2004). The cytoplasmic domain of tight junctions is linked to adaptor proteins including zona-occludins and cingulin (Bazzoni et al. 2000). Similar to adherens junctions, interaction among the transmembrane and cytoplasmic components of the tight junctions and the actin cytoskeleton regulates the strength of the microvessel barrier (Dejana 2004).

3.3.3 Actin Cytoskeleton

As can be evinced from the above discussion, the integrity and stability of both adherens and tight junctions is modulated by the actin cytoskeleton. In endothelial monolayers, stability of the junctional protein complexes is concomitant with the presence of cortical actin (Gulino-Debrac 2013). Disruption of the cortical actin layer in tandem with an increase in stress fiber formation increases stresses on the junctional complexes, resulting in their disassembly (Schnoor et al. 2017). The link between the junctional complexes and actin cytoskeleton is mediated mainly by actin-binding proteins zona occludin-1 and α-catenin, for tight and adherens junctions, respectively (Garcia-Ponce et al. 2015). The specific mechanisms via with these binding proteins function is still under much debate. The debate is much more rigorous since a classic study by Yamada et al. dented a long standing paradigm that α-catenin interacted simultaneously with cadherin and actin (Yamada et al. 2005). It is now emerging that the interaction may depend on the specific form of α-catenin involved (Desai et al. 2013). The complexity of making these determinations exemplifies the dynamic nature of the interactions between junctions and the endothelial cytoskeletal framework. In addition to α-catenin and zona occludin-1, a number of other actin-binding proteins, including vinculin and epithelial protein lost in neoplasm, modulate the actin-junctional complex interactions. For a detailed analysis of the mechanisms associated with these proteins, the reader is referred to other comprehensive reviews on this subject (Oldenburg and de Rooij 2014; Garcia-Ponce et al. 2015).

3.3.4 Small GTPases

Contractility of the actin cytoskeleton in itself is modulated by the Rho family of small GTPases (Duluc and Wojciak-Stothard 2014). In addition to the Rho family, the superfamily of small GTPases includes Rab and Ras families (Takai et al. 2001), which too play a role in mediating endothelial permeability. Overexpression of Rab4 and Rab9, members of the Rab family involved in the regulation of intracellular vesicle trafficking, blunt endocytic uptake of VE-cadherin and thus disruption of the pulmonary endothelial barrier (Chichger et al. 2016). In contrast to Rab-4 and -9, knockdown of Rab11a protects against VE-cadherin recycling, and thus, limits fluid leak across pulmonary microvessels (Yan et al. 2016). Thus, the functional contribution of Rab GTPases to barrier enhancement or disruption appears to be heterogeneous among members of the family. Knockdown of R-Ras, a member of RasGTPases, also increases endothelial barrier permeability (Ichimiya et al. 2015; Vahatupa et al. 2016). In contrast to Rab- and Ras-GTPases, more detailed information is available on the role of RhoGTPases, likely due to their direct impact on the actin cytoskeletal dynamics (Wojciak-Stothard and Leiper 2008; Duluc and Wojciak-Stothard 2014). The Rho/Rac/CDC42 subfamily of the RhoGTPases act as intermediaries for a number of proinflammatory stimuli that disrupt the pulmonary vascular barrier. RhoA and its effector protein Rho kinase are activated by proinflammatory agents, such as histamine and lysophosphatidic acid (Mikelis et al. 2015; Cai et al. 2017). In contrast to Rho, activation of Rac and CDC42 blunt increases in pulmonary endothelial permeability. Rac1 mediates the endothelial barrier enhancing effects of sphingosine-1-phosphate (Singleton et al. 2005; Schlegel and Waschke 2014), and a reduction in Rac1 activity is associated with asymmetric methylarginine-induced pulmonary endothelial barrier breakdown (Wojciak-Stothard et al. 2009). Thus, a complex interplay among the small GTPases in lung endothelial cells allows for the dynamic regulation of the actin cytoskeleton and, thus, plays a role in both pulmonary endothelial barrier disruption and post-injury repair of the barrier.

In addition to RhoGTPases, cytosolic calcium increases via endoplasmic store calcium release could also modulate actin cytoskeleton dynamics. Induction of actin stress fibers in endothelial monolayers in response to thrombin and other proinflammatory agents that increase cytosolic calcium levels is well described (Park et al. 1999; Chiang et al. 2009; Shinde et al. 2013; Absi et al. 2014; Parker et al. 2015). Recent findings suggest that the responses to thrombin treatment are similar in in situ lung microvessels (Escue et al. 2017). Herein, the data revealed that thrombin infusions into microvessels increased endothelial F-actin levels, which in turn was dependent on the release of calcium from endoplasmic stores. However, direct evidence of the mechanisms that tie increased F-actin levels with calcium increase induced by store calcium release is lacking and needs to be elucidated. One possibility is to explore the role of heat shock protein 27, which regulates the actin cytoskeleton by stabilizing the cortical actin and limiting stress fiber formation (Piotrowicz and Levin 1997; Mounier and Arrigo 2002; Sawada et al. 2015). Inactivation of heat shock protein 27 by p38 mitogen-activated protein kinase in a calcium-dependent manner (Huot et al. 1997; An et al. 2005; Evans and Zhao 2017) could be a likely mechanism for actin stress fiber formation in response to proinflammatory mediators.

3.3.5 Gap Junctions

Gap junctions, the third interendothelial junctional complex, act as channels between adjacent cells (Kumar and Gilula 1996; Saez et al. 2003). These channels facilitate the movement of small molecules (<1 kDa) between the connected cells. Connexins, the primary components of gap junctions, assemble into hemichannels or connexins on the plasma membrane (Parthasarathi and Quadri 2009). Connexons, which are hexamers of connexins, diffuse along the plasma membrane toward the intercellular junction. Therein, the hemichannels from apposing plasma membranes then dock together to form a functioning channel. Connexin expression is ubiquitous, and accordingly, connexins are involved in wide-ranging roles from development to apoptosis (Cea et al. 2016; Buo et al. 2017; Uzu et al. 2017). In lung microvessels, gap junctions act as conduits for interendothelial transfer of calcium (Parthasarathi et al. 2006). Thus, gap junction-mediated transfer of calcium increases from one vessel to another could induce proinflammatory responses at the destination vessel. While three connexin subtypes, connexin37, connexin40, and connexin43, predominate the pulmonary vasculature, inhibition of connexin43-containing gap junctions completely abrogates the inter-vascular calcium transfer (Parthasarathi et al. 2006). These data suggest the importance of connexin43 in proinflammatory responses in lung microvessels. In addition, endothelial connexin40 has also been shown to facilitate intercellular propagation of hypoxia-induced membrane depolarization, resulting in vasoconstriction of upstream vessels (Wang et al. 2012).

However, emerging evidence suggest that in addition to their conduit function, gap junctions could act as regulators of permeability in microvessels (Parthasarathi 2012; Soon et al. 2016). Increases in microvascular permeability initiated by instillation of hydrochloric acid in adjoining alveoli are reversibly blocked by inhibiting connexin43-dependent intercellular communication using gap peptides specific to connexin43 (Parthasarathi 2012). This blunting of permeability increases is evident not only at the single microvessel level but also in larger regions of a lung subjected to injury via intra-tracheal acid instillation (Parthasarathi and Bhattacharya 2011). An intriguing nature of these findings is that the increased strength of the barrier is synchronous with the loss of gap junction communication. This raises the question on how loss of the conduit function of gap junctions impacts on the barrier protective function of adherens and tight junctions. It is possible that the physical interaction of gap junctions with tight and adherens junction proteins (Nagasawa et al. 2006; Chang et al. 2014; Ambrosi et al. 2016; Radeva and Waschke 2017) could underlie the functional interaction among these junctions. In brain and retinal endothelial cells, it has been posited that physical interaction between zona occludin-1 and connexin43 underlies the reduction in barrier strength when the interendothelial gap junction communication is blocked (Nagasawa et al. 2006; Tien et al. 2013). Thus, studies using endothelial cells of the systemic vasculature support the possibility of physical and functional interaction between gap junctions and other junctional proteins, albeit in a diametrically inverse direction. Loss of gap junction communication weakens the endothelial barrier in the systemic circulation, while it strengthens the barrier in the pulmonary circulation, suggesting that the barrier regulatory mechanisms in the two circulations may be dissimilar. Support for functional interaction between gap and adherens junction in pulmonary microvessels comes from studies to determine the long-term effects of endotoxin. Airway instillation of endotoxin in rats caused an acute increase in microvessel permeability, which declined to baseline levels in 5 days (Kandasamy et al. 2015). In situ immunofluorescence of VE-cadherin and connexin43 expression in lung microvessels following endotoxin treatment revealed that VE-cadherin levels declined in the acute phase and returned to baseline levels, while connexin43 levels increased in the acute phase, but declined to near zero in 5 days (Kandasamy et al. 2015). The interaction between the two proteins is further supported by findings that show an increase in VE-cadherin levels in lung microvessels in response to in vivo knockdown of connexin43, elicited by tail vein injection of connexin43 shRNA. However, the intermediary between the loss of gap junction communication and increase in VE-cadherin expression remains to be delineated.

3.4 Glycocalyx

The luminal side of the vascular endothelium is layered with a variety of macromolecules, together termed the glycocalyx (Collins et al. 2013). The major constituent molecules of the glycocalyx are proteoglycans and glycosaminoglycan, whose composition is continually modified via self-assembly and shedding (Reitsma et al. 2007; Woodcock and Woodcock 2012). The glycocalyx extends into the interendothelial cleft and forms a physical and functional barrier against the movement of proteins across the vasculature (Reitsma et al. 2007; Dull et al. 2012). In systemic vessels, damage to the glycocalyx, as assessed by penetration of dextran molecules into the glycocalyx layer or presence of glycocalyx components in plasma, increases under inflammatory conditions such as sepsis and due to the presence of inflammatory mediators such as TNF-α (Henry and Duling 2000; Steppan et al. 2011), leading to increased microvascular permeability (Salmon and Satchell 2012). However, in the pulmonary microcirculation, evidence of a role for glycocalyx in permeability regulation has been at best, indirect. In human lung endothelial cell monolayers, knockdown of glycocalyx components increases fluid flux across the monolayer (Wu et al. 2017). Enzymatic degradation of sialic acid, a glycocalyx component that confers a protective negative charge, increases endothelial monolayer permeability and alveolar fluid filling in isolated lungs (Cioffi et al. 2012). Inhibiting enzymatic degradation of glycocalyx components by lipopolysaccharide limits the development of pulmonary edema (Wang et al. 2017). Thus, while these studies suggest the possibility that the glycocalyx may be involved in regulating lung microvessel permeability, additional studies are needed to strengthen this hypothesis and shed more light on the specific mechanisms and pathways involved.

3.5 Pericytes

Pericytes are located on the abluminal side of pulmonary microvessels and maintain close contact with the endothelium (Schallek et al. 2013). Their location and morphology pose difficulties in identifying and studying them in situ. However, a recent consensus has emerged that cells in close proximity to the endothelium and those that show reactivity to two markers, platelet-dependent growth factor receptor-β and nerve-glial antigen 2, could be classified as pericytes (Navarro et al. 2016). This newer definition has facilitated pericyte identification and characterization. The proposal that pericytes play a role in regulating fluid and solute flux across lung microvessels was presented more than 20 years ago (Lonigro et al. 1996). The ability of lung pericytes in culture to contract in the presence of certain inflammatory agents (Speyer et al. 1999; Donoghue et al. 2006; Kerkar et al. 2006), suggest that they may play an active, rather than passive role in permeability regulation in the lung. However, in vivo evidence that directly support a role for pericytes in pulmonary microvessel permeability remains elusive. Indirect evidence in this regard include recent data showing that lipopolysaccharide treatment results in loss of pericytes, microvessel dysfunction, and attendant increase in fluid leak in mice (Zeng et al. 2016). Further, experiments using endothelial-pericyte coculture in a three-dimensional culture model reveal that the presence of pericytes enhances endothelial junctional protein coupling and reduction in permeability (Bichsel et al. 2015). In contrast to these studies, recent findings that ablation of pericyte-like cells in mouse lungs using diphtheria toxin does not modify the protein content in bronchoalveolar lavage (Hung et al. 2017a), oppose the notion that pericytes are involved in permeability regulation in lungs. These divergent findings could be due to the emerging role of pericytes as active mediators of inflammatory responses. Accordingly, in response to lipopolysaccharide treatment, pericytes in murine lungs have been shown to secrete chemokines (Hung et al. 2017b). Hence, it is possible that pericytes play a more active role in an inflammatory milieu. Additional novel approaches are thus needed to specifically address the role pericytes in lung microvessel permeability.

3.6 Microvessel Heterogeneity

To fully appreciate the signaling mechanisms relevant to the pulmonary endothelial barrier that have been elucidated thus far, two caveats need to be borne in mind. One is the heterogeneous nature of the vascular endothelium, both in terms of the molecular properties and function. In a broad sense, it is well established that significant differences exist between the pulmonary, systemic, and lymphatic vasculature (Potente and Makinen 2017). Juxtaposing characteristics of the vasculatures include hypoxia-induced changes in vascular dimensions, sites of leukocyte retention and migration, and proteoglycan expression (McMahon et al. 2002;

Paffett and Walker 2007; Calabrese et al. 2011; Gane and Stockley 2012). It has been shown recently that inflammatory stimuli elicit heterogeneous responses in different vascular beds (Scott et al. 2013). Even within the pulmonary vasculature, a great degree of phenotypic heterogeneity is evident (Stevens 2011). Expression of proteins including, endothelial nitric oxide synthase, nucleosome assembly protein-1, transient receptor potential channels, and Weibel–Palade bodies differ between pulmonary arteries and microvessels (Stevens 2005; Clark et al. 2008; Cioffi et al. 2009; Ochoa et al. 2010). Further, vascular barrier strength differs between extra-alveolar vessels and septal capillaries (Lowe et al. 2007). Interestingly, endothelial heterogeneity extends even to single pulmonary microvessels. Within single post-capillary venules, some endothelial cells may act as pacemaker cells (Ying et al. 1996). The pacemakers may have a greater number of organelles and thus, exhibit exaggerated responses to inflammatory stimuli in comparison to their neighboring endothelial cells (Parthasarathi et al. 2002; Parthasarathi 2012). In addition to pulmonary microvessels, pulmonary arteries too exhibit regional differences in reactivity (Stack et al. 2016). The extensive inter- and intra-vascular heterogeneity behooves us to limit broad generalizations of data on signaling and mechanisms captured using vessels or cells from a single pulmonary bed. By the same token, the heterogeneous nature of the pulmonary vasculature needs to be implied when extending molecular and functional properties specific to a subsection of the pulmonary vasculature to other subsections.

3.7 Endothelial Monolayers Versus Intact Lung Models

A second caveat involves the differences between data derived using pulmonary endothelial cells and vessels in situ. Due to methodological difficulties associated with characterizing signaling mechanisms in lung vasculature in situ, the bulk of the reported studies are based on monolayers of endothelial cells isolated from pulmonary arteries or microvessels. Alternatives to monolayer-based elucidations include the isolated blood-perfused lung preparation and surgically implanted thoracic window preparation, wherein the endothelial cell mechanisms can be observed in situ (Tabuchi and Kuebler 2008; Parthasarathi 2012; Kandasamy and Parthasarathi 2014). Though these alternatives do have their own limitations, they offer the added advantage of being able to specify the specific subsection of the pulmonary vasculature from which the data is obtained and thus limit errors due to endothelial heterogeneity (Kandasamy and Parthasarathi 2014). A second advantage of the alternative strategies is that the endothelial cells remain in their native environment. The disadvantages of the intact lung preparations are that they are complex and expensive, which likely limits their use. However, though widely utilized, growing evidence indicate that signaling mechanisms delineated using endothelial cells grown in monolayers differ significantly from those defined using in situ endothelial cells (Kandasamy et al. 2013; Uhlig et al. 2014; Aman et al. 2016). In addition to endothelial heterogeneity, a possible reason for the differences in signaling could be that endothelial cells in situ interact with other cells and the matrix within their microenvironment (Aman et al. 2016), a feature that is not available for cells in culture. These differences raise the question on how representative are the data derived using endothelial monolayers. Our studies using isolated lung preparations show that the overall signaling pathways defined in single microvessels concur with that from endothelial monolayers. However, the functional, temporal, and spatial aspects of the signaling are different between the two systems (Kandasamy et al. 2013; Escue et al. 2017). To mitigate these issues, alternate experimental approaches are needed for studies on pulmonary endothelial signaling. Recent advances, such as endothelial–epithelial cocultures, transwell culture methods, and microfluidic platforms (Cao et al. 2016; Chonan et al. 2017; Kim et al. 2017), may well provide a middle path that facilitates generation of more physiologically accurate data.

3.8 Conclusion

Dysregulated increases in pulmonary microvascular permeability contribute to reduced quality of life and increased mortality in affected patients. In addition to microvessels, pathological increases in permeability also occur in pulmonary arteries, which could play a role in pulmonary artery hypertension (Zhou et al. 2016; Satoh et al. 2017). Thus, blunting and eventually reversing pathophysiological disruptions of the pulmonary vascular barrier is key step to restoring normal lung function. Toward this end, several potential targets have been proposed, though their success in mitigating lung injury in clinical practice remains elusive (Bosma et al. 2010; Confalonieri et al. 2017). Pharmacological therapies remain unsuccessful (Matthay et al. 2017) and could be due to the possibility that the underlying causes of endothelial barrier dysfunction are heterogeneous. Intriguingly, several of the proposed therapeutic possibilities have been proven to successfully mitigate vascular dysfunction in animal and cell-based models and yet have failed to yield similar success in clinical trials (Network 2000; Jerng et al. 2006; Rice et al. 2011; Schwartz et al. 2015; Evans and Zhao 2017; Nagendran et al. 2017). Thus, newer therapies based on novel approaches, which take into account pulmonary vascular heterogeneity discussed above, are critically needed. In this regard, use of mesenchymal stem cells toward repairing the endothelial barrier shows promise with success in preclinical trials (Huppert and Matthay 2017). Similar novel and well-designed approaches could well serve to improve quality of life and increase survival in patients afflicted by pulmonary microvascular dysfunction.

Acknowledgement The author sincerely thanks Ms. Gayathri Kaushik for the extensive editorial assistance toward preparation of this manuscript.

References

Absi M, Bruce JI, Ward DT (2014) The inhibitory effect of simvastatin and aspirin on histamine responsiveness in human vascular endothelial cells. Am J Physiol Cell Physiol 306:C679–C686. https://doi.org/10.1152/ajpcell.00304.2013

Adam AP (2015) Regulation of endothelial adherens junctions by tyrosine phosphorylation. Mediators Inflamm 2015:272858. https://doi.org/10.1155/2015/272858

Ahmmed GU, Malik AB (2005) Functional role of TRPC channels in the regulation of endothelial permeability. Pflugers Arch 451:131–142. https://doi.org/10.1007/s00424-005-1461-z

Aman J, Weijers EM, van Nieuw Amerongen GP, Malik AB, van Hinsbergh VW (2016) Using cultured endothelial cells to study endothelial barrier dysfunction: challenges and opportunities. Am J Physiol Lung Cell Mol Physiol 311:L453–L466. https://doi.org/10.1152/ajplung.00393.2015

Ambrosi C, Ren C, Spagnol G et al (2016) Connexin43 forms supramolecular complexes through non-overlapping binding sites for drebrin, tubulin, and ZO-1. PLoS One 11:e0157073. https://doi.org/10.1371/journal.pone.0157073

Ambudkar IS, de Souza LB, Ong HL (2017) TRPC1, Orai1, and STIM1 in SOCE: friends in tight spaces. Cell Calcium 63:33–39. https://doi.org/10.1016/j.ceca.2016.12.009

An SS, Pennella CM, Gonnabathula A et al (2005) Hypoxia alters biophysical properties of endothelial cells via p38 MAPK- and Rho kinase-dependent pathways. Am J Physiol Cell Physiol 289:C521–C530. https://doi.org/10.1152/ajpcell.00429.2004

Armstrong SM, Khajoee V, Wang C et al (2012) Co-regulation of transcellular and paracellular leak across microvascular endothelium by dynamin and Rac. Am J Pathol 180:1308–1323. https://doi.org/10.1016/j.ajpath.2011.12.002

Bai C, Fukuda N, Song Y, Ma T, Matthay MA, Verkman AS (1999) Lung fluid transport in aquaporin-1 and aquaporin-4 knockout mice. J Clin Invest 103:555–561. https://doi.org/10.1172/JCI4138

Bazzoni G, Martinez-Estrada OM, Orsenigo F, Cordenonsi M, Citi S, Dejana E (2000) Interaction of junctional adhesion molecule with the tight junction components ZO-1, cingulin, and occludin. J Biol Chem 275:20520–20526. https://doi.org/10.1074/jbc.M905251199

Bichsel CA, Hall SR, Schmid RA, Guenat OT, Geiser T (2015) Primary human lung pericytes support and stabilize in vitro perfusable microvessels. Tissue Eng Part A 21:2166–2176. https://doi.org/10.1089/ten.TEA.2014.0545

Bijli KM, Fazal F, Slavin SA et al (2016) Phospholipase C-epsilon signaling mediates endothelial cell inflammation and barrier disruption in acute lung injury. Am J Physiol Lung Cell Mol Physiol 311:L517–L524. https://doi.org/10.1152/ajplung.00069.2016

Bosma KJ, Taneja R, Lewis JF (2010) Pharmacotherapy for prevention and treatment of acute respiratory distress syndrome: current and experimental approaches. Drugs 70:1255–1282. https://doi.org/10.2165/10898570-000000000-00000

Buo AM, Tomlinson RE, Eidelman ER, Chason M, Stains JP (2017) Connexin43 and Runx2 interact to affect cortical bone geometry, skeletal development, and osteoblast and osteoclast function. J Bone Miner Res. https://doi.org/10.1002/jbmr.3152

Cai J, Wei J, Li S, Suber T, Zhao J (2017) AM966, an antagonist of lysophosphatidic acid receptor 1, increases lung microvascular endothelial permeability through activation of Rho signaling pathway and phosphorylation of VE-cadherin. Mediators Inflamm 2017:6893560. https://doi.org/10.1155/2017/6893560

Calabrese GC, Gazzaniga S, Oberkersch R, Wainstok R (2011) Decorin and biglycan expression: its relation with endothelial heterogeneity. Histol Histopathol 26:481–490. 10.14670/HH-26.481

Cao Y, Roursgaard M, Jacobsen NR, Moller P, Loft S (2016) Monocyte adhesion induced by multi-walled carbon nanotubes and palmitic acid in endothelial cells and alveolar-endothelial co-cultures. Nanotoxicology 10:235–244. https://doi.org/10.3109/17435390.2015.1048325

Cea LA, Puebla C, Cisterna BA et al (2016) Fast skeletal myofibers of mdx mouse, model of Duchenne muscular dystrophy, express connexin hemichannels that lead to apoptosis. Cell Mol Life Sci 73:2583–2599. https://doi.org/10.1007/s00018-016-2132-2

Chang SF, Chen LJ, Lee PL, Lee DY, Chien S, Chiu JJ (2014) Different modes of endothelial-smooth muscle cell interaction elicit differential beta-catenin phosphorylations and endothelial functions. Proc Natl Acad Sci U S A 111:1855–1860. https://doi.org/10.1073/pnas.1323761111

Chiang ET, Camp SM, Dudek SM et al (2009) Protective effects of high-molecular weight polyethylene glycol (PEG) in human lung endothelial cell barrier regulation: role of actin cytoskeletal rearrangement. Microvasc Res 77:174–186. https://doi.org/10.1016/j.mvr.2008.11.007

Chichger H, Braza J, Duong H, Boni G, Harrington EO (2016) Select Rab GTPases regulate the pulmonary endothelium via endosomal trafficking of vascular endothelial-cadherin. Am J Respir Cell Mol Biol 54:769–781. https://doi.org/10.1165/rcmb.2015-0286OC

Chonan Y, Taki S, Sampetrean O, Saya H, Sudo R (2017) Endothelium-induced three-dimensional invasion of heterogeneous glioma initiating cells in a microfluidic coculture platform. Integr Biol (Camb). https://doi.org/10.1039/c7ib00091j

Cioffi DL, Lowe K, Alvarez DF, Barry C, Stevens T (2009) TRPing on the lung endothelium: calcium channels that regulate barrier function. Antioxid Redox Signal 11:765–776. https://doi.org/10.1089/ARS.2008.2221

Cioffi DL, Pandey S, Alvarez DF, Cioffi EA (2012) Terminal sialic acids are an important determinant of pulmonary endothelial barrier integrity. Am J Physiol Lung Cell Mol Physiol 302:L1067–L1077. https://doi.org/10.1152/ajplung.00190.2011

Clark J, Alvarez DF, Alexeyev M, King JA, Huang L, Yoder MC, Stevens T (2008) Regulatory role for nucleosome assembly protein-1 in the proliferative and vasculogenic phenotype of pulmonary endothelium. Am J Physiol Lung Cell Mol Physiol 294:L431–L439. https://doi.org/10.1152/ajplung.00316.2007

Collins SR, Blank RS, Deatherage LS, Dull RO (2013) Special article: The endothelial glycocalyx: emerging concepts in pulmonary edema and acute lung injury. Anesth Analg 117:664–674. https://doi.org/10.1213/ANE.0b013e3182975b85

Confalonieri M, Salton F, Fabiano F (2017) Acute respiratory distress syndrome. Eur Respir Rev 26. https://doi.org/10.1183/16000617.0116-2016

Corada M, Zanetta L, Orsenigo F et al (2002) A monoclonal antibody to vascular endothelial-cadherin inhibits tumor angiogenesis without side effects on endothelial permeability. Blood 100:905–911

Cosentini R, Aliberti S, Bignamini A, Piffer F, Brambilla AM (2009) Mortality in acute cardiogenic pulmonary edema treated with continuous positive airway pressure. Intensive Care Med 35:299–305. https://doi.org/10.1007/s00134-008-1281-7

Dejana E (2004) Endothelial cell-cell junctions: happy together. Nat Rev Mol Cell Biol 5:261–270. https://doi.org/10.1038/nrm1357

Dejana E, Orsenigo F (2013) Endothelial adherens junctions at a glance. J Cell Sci 126:2545–2549. https://doi.org/10.1242/jcs.124529

Dejana E, Orsenigo F, Lampugnani MG (2008) The role of adherens junctions and VE-cadherin in the control of vascular permeability. J Cell Sci 121:2115–2122. https://doi.org/10.1242/jcs.017897

Desai R, Sarpal R, Ishiyama N, Pellikka M, Ikura M, Tepass U (2013) Monomeric alpha-catenin links cadherin to the actin cytoskeleton. Nat Cell Biol 15:261–273. https://doi.org/10.1038/ncb2685

Donoghue L, Tyburski JG, Steffes CP, Wilson RF (2006) Vascular endothelial growth factor modulates contractile response in microvascular lung pericytes. Am J Surg 191:349–352. https://doi.org/10.1016/j.amjsurg.2005.10.034

Dorland YL, Huveneers S (2017) Cell-cell junctional mechanotransduction in endothelial remodeling. Cell Mol Life Sci 74:279–292. https://doi.org/10.1007/s00018-016-2325-8

Drab M, Verkade P, Elger M et al (2001) Loss of caveolae, vascular dysfunction, and pulmonary defects in caveolin-1 gene-disrupted mice. Science 293:2449–2452. https://doi.org/10.1126/science.1062688

Dull RO, Cluff M, Kingston J et al (2012) Lung heparan sulfates modulate K(fc) during increased vascular pressure: evidence for glycocalyx-mediated mechanotransduction. Am J Physiol Lung Cell Mol Physiol 302:L816–L828. https://doi.org/10.1152/ajplung.00080.2011

Duluc L, Wojciak-Stothard B (2014) Rho GTPases in the regulation of pulmonary vascular barrier function. Cell Tissue Res 355:675–685. https://doi.org/10.1007/s00441-014-1805-0

Eltzschig HK, Carmeliet P (2011) Hypoxia and inflammation. N Engl J Med 364:656–665. https://doi.org/10.1056/NEJMra0910283

Eltzschig HK, Bratton DL, Colgan SP (2014) Targeting hypoxia signalling for the treatment of ischaemic and inflammatory diseases. Nat Rev Drug Discov 13:852–869. https://doi.org/10.1038/nrd4422

Escue R, Kandasamy K, Parthasarathi K (2017) Thrombin induces inositol trisphosphate-mediated spatially extensive responses in lung microvessels. Am J Pathol 187:921–935. https://doi.org/10.1016/j.ajpath.2016.12.014

Evans CE, Zhao YY (2017) Impact of thrombosis on pulmonary endothelial injury and repair following sepsis. Am J Physiol Lung Cell Mol Physiol 312:L441–L451. https://doi.org/10.1152/ajplung.00441.2016

Fainaru O, Adini I, Benny O, Bazinet L, Pravda E, D'Amato R, Folkman J (2008) Doxycycline induces membrane expression of VE-cadherin on endothelial cells and prevents vascular hyperpermeability. FASEB J 22:3728–3735. https://doi.org/10.1096/fj.08-110494

Gane J, Stockley R (2012) Mechanisms of neutrophil transmigration across the vascular endothelium in COPD. Thorax 67:553–561. https://doi.org/10.1136/thoraxjnl-2011-200088

Gao X, Kouklis P, Xu N, Minshall RD, Sandoval R, Vogel SM, Malik AB (2000) Reversibility of increased microvessel permeability in response to VE-cadherin disassembly. Am J Physiol Lung Cell Mol Physiol 279:L1218–L1225

Garcia-Ponce A, Citalan-Madrid AF, Velazquez-Avila M, Vargas-Robles H, Schnoor M (2015) The role of actin-binding proteins in the control of endothelial barrier integrity. Thromb Haemost 113:20–36. https://doi.org/10.1160/TH14-04-0298

Gavard J (2014) Endothelial permeability and VE-cadherin: a wacky comradeship. Cell Adh Migr 8:158–164

Geyer M, Huang F, Sun Y, Vogel SM, Malik AB, Taylor CW, Komarova YA (2015) Microtubule-associated protein EB3 regulates IP3 receptor clustering and Ca(2+) signaling in endothelial cells. Cell Rep 12:79–89. https://doi.org/10.1016/j.celrep.2015.06.001

Gonzales JN, Lucas R, Verin AD (2015) The acute respiratory distress syndrome: mechanisms and perspective therapeutic approaches. Austin J Vasc Med 2(1)

Gulino-Debrac D (2013) Mechanotransduction at the basis of endothelial barrier function. Tissue Barriers 1:e24180. https://doi.org/10.4161/tisb.24180

Hansen CG, Nichols BJ (2010) Exploring the caves: cavins, caveolins and caveolae. Trends Cell Biol 20:177–186. https://doi.org/10.1016/j.tcb.2010.01.005

Hansen CG, Shvets E, Howard G, Riento K, Nichols BJ (2013) Deletion of cavin genes reveals tissue-specific mechanisms for morphogenesis of endothelial caveolae. Nat Commun 4:1831. https://doi.org/10.1038/ncomms2808

Heimerl S, Liebisch G, Le Lay S et al (2008) Caveolin-1 deficiency alters plasma lipid and lipoprotein profiles in mice. Biochem Biophys Res Commun 367:826–833. https://doi.org/10.1016/j.bbrc.2008.01.010

Henry CB, Duling BR (2000) TNF-alpha increases entry of macromolecules into luminal endothelial cell glycocalyx. Am J Physiol Heart Circ Physiol 279:H2815–H2823

Herwig MC, Muller KM, Muller AM (2008) Endothelial VE-cadherin expression in human lungs. Pathol Res Pract 204:725–730. https://doi.org/10.1016/j.prp.2008.04.014

Huang Y, He Q (2017) Inhibition of c-Src protects paraquat induced microvascular endothelial injury by modulating caveolin-1 phosphorylation and caveolae mediated transcellular permeability. Environ Toxicol Pharmacol 52:62–68. https://doi.org/10.1016/j.etap.2017.01.023

Hung CF, Chow YH, Liles WC, Altemeier WA, Schnapp LM (2017a) Ablation of pericyte-like cells in lungs by oropharyngeal aspiration of diphtheria toxin. Am J Respir Cell Mol Biol 56:160–167. https://doi.org/10.1165/rcmb.2016-0083MA

Hung CF, Mittelsteadt KL, Brauer R et al (2017b) Lung pericyte-like cells are functional interstitial immune sentinel cells. Am J Physiol Lung Cell Mol Physiol 312:L556–L567. https://doi.org/10.1152/ajplung.00349.2016

Huot J, Houle F, Marceau F, Landry J (1997) Oxidative stress-induced actin reorganization mediated by the p38 mitogen-activated protein kinase/heat shock protein 27 pathway in vascular endothelial cells. Circ Res 80:383–392

Huppert LA, Matthay MA (2017) Alveolar fluid clearance in pathologically relevant conditions: in vitro and in vivo models of acute respiratory distress syndrome. Front Immunol 8:371. https://doi.org/10.3389/fimmu.2017.00371

Ichimiya H, Maeda K, Enomoto A, Weng L, Takahashi M, Murohara T (2015) Girdin/GIV regulates transendothelial permeability by controlling VE-cadherin trafficking through the small GTPase, R-Ras. Biochem Biophys Res Commun 461:260–267. https://doi.org/10.1016/j.bbrc.2015.04.012

Ieguchi K, Omori T, Komatsu A, Tomita T, Deguchi A, Maru Y (2013) Ephrin-A1 expression induced by S100A8 is mediated by the toll-like receptor 4. Biochem Biophys Res Commun 440:623–629. https://doi.org/10.1016/j.bbrc.2013.09.119

Iyer S, Ferreri DM, DeCocco NC, Minnear FL, Vincent PA (2004) VE-cadherin-p120 interaction is required for maintenance of endothelial barrier function. Am J Physiol Lung Cell Mol Physiol 286:L1143–L1153. https://doi.org/10.1152/ajplung.00305.2003

Jerng JS, Yu CJ, Wang HC, Chen KY, Cheng SL, Yang PC (2006) Polymorphism of the angiotensin-converting enzyme gene affects the outcome of acute respiratory distress syndrome. Crit Care Med 34:1001–1006. https://doi.org/10.1097/01.CCM.0000206107.92476.39

Kandasamy K, Parthasarathi K (2014) Quantifying single microvessel permeability in isolated blood-perfused rat lung preparation. J Vis Exp:e51552. https://doi.org/10.3791/51552

Kandasamy K, Bezavada L, Escue RB, Parthasarathi K (2013) Lipopolysaccharide induces endoplasmic store Ca2+-dependent inflammatory responses in lung microvessels. PLoS One 8:e63465. https://doi.org/10.1371/journal.pone.0063465

Kandasamy K, Escue R, Manna J, Adebiyi A, Parthasarathi K (2015) Changes in endothelial connexin 43 expression inversely correlate with microvessel permeability and VE-cadherin expression in endotoxin-challenged lungs. Am J Physiol Lung Cell Mol Physiol 309:L584–L592. https://doi.org/10.1152/ajplung.00211.2014

Kerkar S, Williams M, Blocksom JM, Wilson RF, Tyburski JG, Steffes CP (2006) TNF-alpha and IL-1beta increase pericyte/endothelial cell co-culture permeability. J Surg Res 132:40–45. https://doi.org/10.1016/j.jss.2005.06.033

Kim HR, Cho HS, Shin DY, Chung KH (2017) Novel approach to study the cardiovascular effects and mechanism of action of urban particulate matter using lung epithelial-endothelial tetra-culture system. Toxicol In Vitro 38:33–40. https://doi.org/10.1016/j.tiv.2016.11.001

King LS, Nielsen S, Agre P, Brown RH (2002) Decreased pulmonary vascular permeability in aquaporin-1-null humans. Proc Natl Acad Sci U S A 99:1059–1063. https://doi.org/10.1073/pnas.022626499

Kuebler WM, Wittenberg C, Lee WL et al (2016) Thrombin stimulates albumin transcytosis in lung microvascular endothelial cells via activation of acid sphingomyelinase. Am J Physiol Lung Cell Mol Physiol 310:L720–L732. https://doi.org/10.1152/ajplung.00157.2015

Kumar NM, Gilula NB (1996) The gap junction communication channel. Cell 84:381–388

Le Lay S, Kurzchalia TV (2005) Getting rid of caveolins: phenotypes of caveolin-deficient animals. Biochim Biophys Acta 1746:322–333. https://doi.org/10.1016/j.bbamcr.2005.06.001

Li HH, Li J, Wasserloos KJ et al (2013) Caveolae-dependent and -independent uptake of albumin in cultured rodent pulmonary endothelial cells. PLoS One 8:e81903. https://doi.org/10.1371/journal.pone.0081903

Lonigro AJ, McMurdo L, Stephenson AH, Sprague RS, Weintraub NL (1996) Hypotheses regarding the role of pericytes in regulating movement of fluid, nutrients, and hormones across the microcirculatory endothelial barrier. Diabetes 45(Suppl 1):S38–S43

Lowe K, Alvarez D, King J, Stevens T (2007) Phenotypic heterogeneity in lung capillary and extra-alveolar endothelial cells. Increased extra-alveolar endothelial permeability is sufficient to decrease compliance. J Surg Res 143:70–77. https://doi.org/10.1016/j.jss.2007.03.047

Lucas R, Yang G, Gorshkov BA et al (2012) Protein kinase C-alpha and arginase I mediate pneumolysin-induced pulmonary endothelial hyperpermeability. Am J Respir Cell Mol Biol 47:445–453. https://doi.org/10.1165/rcmb.2011-0332OC

Malczyk M, Erb A, Veith C et al (2017) The role of transient receptor potential channel 6 channels in the pulmonary vasculature. Front Immunol 8:707. https://doi.org/10.3389/fimmu.2017.00707

Maniatis NA, Chernaya O, Shinin V, Minshall RD (2012) Caveolins and lung function. Adv Exp Med Biol 729:157–179. https://doi.org/10.1007/978-1-4614-1222-9_11

Matthay MA, Ware LB, Zimmerman GA (2012) The acute respiratory distress syndrome. J Clin Invest 122:2731–2740. https://doi.org/10.1172/JCI60331

Matthay MA, McAuley DF, Ware LB (2017) Clinical trials in acute respiratory distress syndrome: challenges and opportunities. Lancet Respir Med 5:524–534. https://doi.org/10.1016/S2213-2600(17)30188-1

McMahon TJ, Moon RE, Luschinger BP et al (2002) Nitric oxide in the human respiratory cycle. Nat Med 8:711–717. https://doi.org/10.1038/nm718

Mikelis CM, Simaan M, Ando K et al (2015) RhoA and ROCK mediate histamine-induced vascular leakage and anaphylactic shock. Nat Commun 6:6725. https://doi.org/10.1038/ncomms7725

Milici AJ, Watrous NE, Stukenbrok H, Palade GE (1987) Transcytosis of albumin in capillary endothelium. J Cell Biol 105:2603–2612

Mounier N, Arrigo AP (2002) Actin cytoskeleton and small heat shock proteins: how do they interact? Cell Stress Chaperones 7:167–176

Muramatsu F, Kidoya H, Naito H, Hayashi Y, Iba T, Takakura N (2017) Plakoglobin maintains the integrity of vascular endothelial cell junctions and regulates VEGF-induced phosphorylation of VE-cadherin. J Biochem doi. https://doi.org/10.1093/jb/mvx001

Nagasawa K, Chiba H, Fujita H, Kojima T, Saito T, Endo T, Sawada N (2006) Possible involvement of gap junctions in the barrier function of tight junctions of brain and lung endothelial cells. J Cell Physiol 208:123–132. https://doi.org/10.1002/jcp.20647

Nagendran M, McAuley DF, Kruger PS et al (2017) Statin therapy for acute respiratory distress syndrome: an individual patient data meta-analysis of randomised clinical trials. Intensive Care Med 43:663–671. https://doi.org/10.1007/s00134-016-4649-0

Navarro R, Compte M, Alvarez-Vallina L, Sanz L (2016) Immune regulation by pericytes: modulating innate and adaptive immunity. Front Immunol 7:480. https://doi.org/10.3389/fimmu.2016.00480

Network TANAftA (2000) Ketoconazole for early treatment of acute lung injury and acute respiratory distress syndrome: a randomized controlled trial. The ARDS Network. JAMA 283:1995–2002

Ochoa CD, Wu S, Stevens T (2010) New developments in lung endothelial heterogeneity: Von Willebrand factor, P-selectin, and the Weibel-Palade body. Semin Thromb Hemost 36:301–308. https://doi.org/10.1055/s-0030-1253452

Oldenburg J, de Rooij J (2014) Mechanical control of the endothelial barrier. Cell Tissue Res 355:545–555. https://doi.org/10.1007/s00441-013-1792-6

Orsenigo F, Giampietro C, Ferrari A et al (2012) Phosphorylation of VE-cadherin is modulated by haemodynamic forces and contributes to the regulation of vascular permeability in vivo. Nat Commun 3:1208. https://doi.org/10.1038/ncomms2199

Paffett ML, Walker BR (2007) Vascular adaptations to hypoxia: molecular and cellular mechanisms regulating vascular tone. Essays Biochem 43:105–119. https://doi.org/10.1042/BSE0430105

Palade GE (1953) Fine structure of blood capillaries. J Appl Phys 24:1424

Palade GE (1961) Blood capillaries of the heart and other organs. Circulation 24:368–388
Park JH, Okayama N, Gute D, Krsmanovic A, Battarbee H, Alexander JS (1999) Hypoxia/aglycemia increases endothelial permeability: role of second messengers and cytoskeleton. Am J Physiol 277:C1066–C1074
Parker WH, Qu ZC, May JM (2015) Intracellular ascorbate prevents endothelial barrier permeabilization by thrombin. J Biol Chem 290:21486–21497. https://doi.org/10.1074/jbc.M115.662098
Parthasarathi K (2012) Endothelial connexin43 mediates acid-induced increases in pulmonary microvascular permeability. Am J Physiol Lung Cell Mol Physiol 303:L33–L42. https://doi.org/10.1152/ajplung.00219.2011
Parthasarathi K, Bhattacharya J (2011) Localized acid instillation by a wedged-catheter method reveals a role for vascular gap junctions in spatial expansion of acid injury. Anat Rec (Hoboken) 294:1585–1591. https://doi.org/10.1002/ar.21460
Parthasarathi K, Quadri SK (2009) Cadherins and connexins in pulmonary endothelial function, pp 33–50. https://doi.org/10.1002/9780470747490.ch3
Parthasarathi K, Ichimura H, Quadri S, Issekutz A, Bhattacharya J (2002) Mitochondrial reactive oxygen species regulate spatial profile of proinflammatory responses in lung venular capillaries. J Immunol 169:7078–7086
Parthasarathi K, Ichimura H, Monma E, Lindert J, Quadri S, Issekutz A, Bhattacharya J (2006) Connexin 43 mediates spread of Ca2+-dependent proinflammatory responses in lung capillaries. J Clin Invest 116:2193–2200. https://doi.org/10.1172/JCI26605
Parton RG, del Pozo MA (2013) Caveolae as plasma membrane sensors, protectors and organizers. Nat Rev Mol Cell Biol 14:98–112. https://doi.org/10.1038/nrm3512
Piotrowicz RS, Levin EG (1997) Basolateral membrane-associated 27-kDa heat shock protein and microfilament polymerization. J Biol Chem 272:25920–25927
Pires-Neto RC, Del Carlo Bernardi F, Alves de Araujo P, Mauad T, Dolhnikoff M (2016) The expression of water and ion channels in diffuse alveolar damage is not dependent on DAD etiology. PLoS One 11:e0166184. https://doi.org/10.1371/journal.pone.0166184
Potente M, Makinen T (2017) Vascular heterogeneity and specialization in development and disease. Nat Rev Mol Cell Biol 18:477–494. https://doi.org/10.1038/nrm.2017.36
Quadri SK (2012) Cross talk between focal adhesion kinase and cadherins: role in regulating endothelial barrier function. Microvasc Res 83:3–11. https://doi.org/10.1016/j.mvr.2011.08.001
Radeva MY, Waschke J (2017) Mind the gap: mechanisms regulating the endothelial barrier. Acta Physiol (Oxf). https://doi.org/10.1111/apha.12860
Reitsma S, Slaaf DW, Vink H, van Zandvoort MA, oude Egbrink MG (2007) The endothelial glycocalyx: composition, functions, and visualization. Pflugers Arch 454:345–359. https://doi.org/10.1007/s00424-007-0212-8
Rice TW, Wheeler AP, Thompson BT et al (2011) Enteral omega-3 fatty acid, gamma-linolenic acid, and antioxidant supplementation in acute lung injury. JAMA 306:1574–1581. https://doi.org/10.1001/jama.2011.1435
Richter T, Floetenmeyer M, Ferguson C et al (2008) High-resolution 3D quantitative analysis of caveolar ultrastructure and caveola-cytoskeleton interactions. Traffic 9:893–909. https://doi.org/10.1111/j.1600-0854.2008.00733.x
Saez JC, Berthoud VM, Branes MC, Martinez AD, Beyer EC (2003) Plasma membrane channels formed by connexins: their regulation and functions. Physiol Rev 83:1359–1400. https://doi.org/10.1152/physrev.00007.2003
Salmon AH, Satchell SC (2012) Endothelial glycocalyx dysfunction in disease: albuminuria and increased microvascular permeability. J Pathol 226:562–574. https://doi.org/10.1002/path.3964
Satoh T, Satoh K, Yaoita N et al (2017) Activated TAFI promotes the development of chronic thromboembolic pulmonary hypertension: a possible novel therapeutic target. Circ Res 120:1246–1262. https://doi.org/10.1161/CIRCRESAHA.117.310640

Sawada J, Li F, Komatsu M (2015) R-Ras inhibits VEGF-induced p38MAPK activation and HSP27 phosphorylation in endothelial cells. J Vasc Res 52:347–359. https://doi.org/10.1159/000444526

Sawant DA, Tharakan B, Hunter FA, Smythe WR, Childs EW (2011) Role of beta-catenin in regulating microvascular endothelial cell hyperpermeability. J Trauma 70:481–487; discussion 487–488. https://doi.org/10.1097/TA.0b013e31820b3ed7

Schallek J, Geng Y, Nguyen H, Williams DR (2013) Morphology and topography of retinal pericytes in the living mouse retina using in vivo adaptive optics imaging and ex vivo characterization. Invest Ophthalmol Vis Sci 54:8237–8250. https://doi.org/10.1167/iovs.13-12581

Schlegel N, Waschke J (2014) cAMP with other signaling cues converges on Rac1 to stabilize the endothelial barrier – a signaling pathway compromised in inflammation. Cell Tissue Res 355:587–596. https://doi.org/10.1007/s00441-013-1755-y

Schnittler H, Taha M, Schnittler MO, Taha AA, Lindemann N, Seebach J (2014) Actin filament dynamics and endothelial cell junctions: the Ying and Yang between stabilization and motion. Cell Tissue Res 355:529–543. https://doi.org/10.1007/s00441-014-1856-2

Schnitzer JE, Oh P (1996) Aquaporin-1 in plasma membrane and caveolae provides mercury-sensitive water channels across lung endothelium. Am J Physiol 270:H416–H422

Schnoor M, Garcia Ponce A, Vadillo E, Pelayo R, Rossaint J, Zarbock A (2017) Actin dynamics in the regulation of endothelial barrier functions and neutrophil recruitment during endotoxemia and sepsis. Cell Mol Life Sci 74:1985–1997. https://doi.org/10.1007/s00018-016-2449-x

Schubert W, Frank PG, Woodman SE, Hyogo H, Cohen DE, Chow CW, Lisanti MP (2002) Microvascular hyperpermeability in caveolin-1 (-/-) knock-out mice. Treatment with a specific nitric-oxide synthase inhibitor, L-NAME, restores normal microvascular permeability in Cav-1 null mice. J Biol Chem 277:40091–40098. https://doi.org/10.1074/jbc.M205948200

Schwartz AV, Chen H, Ambrosius WT et al (2015) Effects of TZD use and discontinuation on fracture rates in ACCORD bone study. J Clin Endocrinol Metab 100:4059–4066. https://doi.org/10.1210/jc.2015-1215

Scott DW, Vallejo MO, Patel RP (2013) Heterogenic endothelial responses to inflammation: role for differential N-glycosylation and vascular bed of origin. J Am Heart Assoc 2:e000263. https://doi.org/10.1161/JAHA.113.000263

Shang D, Peng T, Gou S, Li Y, Wu H, Wang C, Yang Z (2016) High mobility group box protein 1 boosts endothelial albumin transcytosis through the RAGE/Src/Caveolin-1 pathway. Sci Rep 6:32180. https://doi.org/10.1038/srep32180

Shinde AV, Motiani RK, Zhang X et al (2013) STIM1 controls endothelial barrier function independently of Orai1 and Ca2+ entry. Sci Signal 6:ra18. https://doi.org/10.1126/scisignal.2003425

Singleton PA, Dudek SM, Chiang ET, Garcia JG (2005) Regulation of sphingosine 1-phosphate-induced endothelial cytoskeletal rearrangement and barrier enhancement by S1P1 receptor, PI3 kinase, Tiam1/Rac1, and alpha-actinin. FASEB J 19:1646–1656. https://doi.org/10.1096/fj.05-3928com

Sinha B, Koster D, Ruez R et al (2011) Cells respond to mechanical stress by rapid disassembly of caveolae. Cell 144:402–413. https://doi.org/10.1016/j.cell.2010.12.031

Soni D, Regmi SC, Wang DM et al (2017) Pyk2 phosphorylation of VE-PTP downstream of STIM1-induced Ca2+ entry regulates disassembly of adherens junctions. Am J Physiol Lung Cell Mol Physiol 312:L1003–L1017. https://doi.org/10.1152/ajplung.00008.2017

Soon AS, Chua JW, Becker DL (2016) Connexins in endothelial barrier function – novel therapeutic targets countering vascular hyperpermeability. Thromb Haemost 116:852–867. https://doi.org/10.1160/TH16-03-0210

Sowa G (2012) Caveolae, caveolins, cavins, and endothelial cell function: new insights. Front Physiol 2:120. https://doi.org/10.3389/fphys.2011.00120

Speyer CL, Steffes CP, Ram JL (1999) Effects of vasoactive mediators on the rat lung pericyte: quantitative analysis of contraction on collagen lattice matrices. Microvasc Res 57:134–143. https://doi.org/10.1006/mvre.1998.2134

Stack A, Derksen FJ, Williams KJ, Robinson NE, Jackson WF (2016) Regional heterogeneity in the reactivity of equine small pulmonary blood vessels. J Appl Physiol (1985) 120:599–607. https://doi.org/10.1152/japplphysiol.00975.2015

Steppan J, Hofer S, Funke B et al (2011) Sepsis and major abdominal surgery lead to flaking of the endothelial glycocalix. J Surg Res 165:136–141. https://doi.org/10.1016/j.jss.2009.04.034

Stevens T (2005) Molecular and cellular determinants of lung endothelial cell heterogeneity. Chest 128:558S–564S. https://doi.org/10.1378/chest.128.6_suppl.558S

Stevens T (2011) Functional and molecular heterogeneity of pulmonary endothelial cells. Proc Am Thorac Soc 8:453–457. https://doi.org/10.1513/pats.201101-004MW

Sundivakkam PC, Freichel M, Singh V et al (2012) The Ca(2+) sensor stromal interaction molecule 1 (STIM1) is necessary and sufficient for the store-operated Ca(2+) entry function of transient receptor potential canonical (TRPC) 1 and 4 channels in endothelial cells. Mol Pharmacol 81:510–526. https://doi.org/10.1124/mol.111.074658

Sverdlov M, Shinin V, Place AT, Castellon M, Minshall RD (2009) Filamin A regulates caveolae internalization and trafficking in endothelial cells. Mol Biol Cell 20:4531–4540. https://doi.org/10.1091/mbc.E08-10-0997

Tabuchi A, Kuebler WM (2008) Endothelium-platelet interactions in inflammatory lung disease. Vascul Pharmacol 49:141–150. https://doi.org/10.1016/j.vph.2008.06.004

Takai Y, Sasaki T, Matozaki T (2001) Small GTP-binding proteins. Physiol Rev 81:153–208

Taveau JC, Dubois M, Le Bihan O et al (2008) Structure of artificial and natural VE-cadherin-based adherens junctions. Biochem Soc Trans 36:189–193. https://doi.org/10.1042/BST0360189

Tien T, Barrette KF, Chronopoulos A, Roy S (2013) Effects of high glucose-induced Cx43 downregulation on occludin and ZO-1 expression and tight junction barrier function in retinal endothelial cells. Invest Ophthalmol Vis Sci 54:6518–6525. https://doi.org/10.1167/iovs.13-11763

Tiruppathi C, Ahmmed GU, Vogel SM, Malik AB (2006) Ca2+ signaling, TRP channels, and endothelial permeability. Microcirculation 13:693–708. https://doi.org/10.1080/10739680600930347

Tiruppathi C, Shimizu J, Miyawaki-Shimizu K et al (2008) Role of NF-kappaB-dependent caveolin-1 expression in the mechanism of increased endothelial permeability induced by lipopolysaccharide. J Biol Chem 283:4210–4218. https://doi.org/10.1074/jbc.M703153200

Tsukita S, Furuse M, Itoh M (2001) Multifunctional strands in tight junctions. Nat Rev Mol Cell Biol 2:285–293. https://doi.org/10.1038/35067088

Uhlig S, Yang Y, Waade J, Wittenberg C, Babendreyer A, Kuebler WM (2014) Differential regulation of lung endothelial permeability in vitro and in situ. Cell Physiol Biochem 34:1–19. https://doi.org/10.1159/000362980

Uzu M, Sato H, Shimizu A, Shibata Y, Ueno K, Hisaka A (2017) Connexin 43 enhances Bax activation via JNK activation in sunitinib-induced apoptosis in mesothelioma cells. J Pharmacol Sci 134:101–107. https://doi.org/10.1016/j.jphs.2017.05.005

Vadasz I, Sznajder JI (2017) Gas exchange disturbances regulate alveolar fluid clearance during acute lung injury. Front Immunol 8:757. https://doi.org/10.3389/fimmu.2017.00757

Vahatupa M, Prince S, Vataja S et al (2016) Lack of R-Ras leads to increased vascular permeability in ischemic retinopathy. Invest Ophthalmol Vis Sci 57:4898–4909. https://doi.org/10.1167/iovs.16-19212

Vandenbroucke St Amant E, Tauseef M, Vogel SM, Gao XP, Mehta D, Komarova YA, Malik AB (2012) PKCalpha activation of p120-catenin serine 879 phospho-switch disassembles VE-cadherin junctions and disrupts vascular integrity. Circ Res 111:739–749. https://doi.org/10.1161/CIRCRESAHA.112.269654

Verkman AS (2007) Role of aquaporins in lung liquid physiology. Respir Physiol Neurobiol 159:324–330. https://doi.org/10.1016/j.resp.2007.02.012

Villalta PC, Townsley MI (2013) Transient receptor potential channels and regulation of lung endothelial permeability. Pulm Circ 3:802–815. https://doi.org/10.1086/674765

Vogel SM, Malik AB (2012) Cytoskeletal dynamics and lung fluid balance. Compr Physiol 2:449–478. https://doi.org/10.1002/cphy.c100006

Wang L, Yin J, Nickles HT et al (2012) Hypoxic pulmonary vasoconstriction requires connexin 40-mediated endothelial signal conduction. J Clin Invest 122:4218–4230. https://doi.org/10.1172/JCI59176

Wang N, Zhang D, Sun G, Zhang H, You Q, Shao M, Yue Y (2015) Lipopolysaccharide-induced caveolin-1 phosphorylation-dependent increase in transcellular permeability precedes the increase in paracellular permeability. Drug Des Devel Ther 9:4965–4977. https://doi.org/10.2147/DDDT.S77646

Wang G, Zhang J, Xu C, Han X, Gao Y, Chen H (2016) Inhibition of SOCs attenuates acute lung injury induced by severe acute pancreatitis in rats and PMVECs injury induced by lipopolysaccharide. Inflammation 39:1049–1058. https://doi.org/10.1007/s10753-016-0335-1

Wang T, Gross C, Desai AA et al (2017) Endothelial cell signaling and ventilator-induced lung injury: molecular mechanisms, genomic analyses, and therapeutic targets. Am J Physiol Lung Cell Mol Physiol 312:L452–L476. https://doi.org/10.1152/ajplung.00231.2016

Ware LB, Fremont RD, Bastarache JA, Calfee CS, Matthay MA (2010) Determining the aetiology of pulmonary oedema by the oedema fluid-to-plasma protein ratio. Eur Respir J 35:331–337. https://doi.org/10.1183/09031936.00098709

Weidenfeld S, Kuebler WM (2017) Cytokine-regulation of Na+-K+-Cl- cotransporter 1 and cystic fibrosis transmembrane conductance regulator-potential role in pulmonary inflammation and edema formation. Front Immunol 8:393. https://doi.org/10.3389/fimmu.2017.00393

Wojciak-Stothard B, Leiper J (2008) Rho GTPases and hypoxia in pulmonary vascular endothelial cells. Methods Enzymol 439:267–283. https://doi.org/10.1016/S0076-6879(07)00420-X

Wojciak-Stothard B, Torondel B, Zhao L, Renne T, Leiper JM (2009) Modulation of Rac1 activity by ADMA/DDAH regulates pulmonary endothelial barrier function. Mol Biol Cell 20:33–42. https://doi.org/10.1091/mbc.E08-04-0395

Woodcock TE, Woodcock TM (2012) Revised Starling equation and the glycocalyx model of transvascular fluid exchange: an improved paradigm for prescribing intravenous fluid therapy. Br J Anaesth 108:384–394. https://doi.org/10.1093/bja/aer515

Wu MM, Covington ED, Lewis RS (2014) Single-molecule analysis of diffusion and trapping of STIM1 and Orai1 at endoplasmic reticulum-plasma membrane junctions. Mol Biol Cell 25:3672–3685. https://doi.org/10.1091/mbc.E14-06-1107

Wu F, Peng Z, Park PW, Kozar RA (2017) Loss of Syndecan-1 abrogates the pulmonary protective phenotype induced by plasma after hemorrhagic shock. Shock. https://doi.org/10.1097/SHK.0000000000000832

Yamada S, Pokutta S, Drees F, Weis WI, Nelson WJ (2005) Deconstructing the cadherin-catenin-actin complex. Cell 123:889–901. https://doi.org/10.1016/j.cell.2005.09.020

Yan Z, Wang ZG, Segev N et al (2016) Rab11a mediates vascular endothelial-cadherin recycling and controls endothelial barrier function. Arterioscler Thromb Vasc Biol 36:339–349. https://doi.org/10.1161/ATVBAHA.115.306549

Yazbeck P, Tauseef M, Kruse K et al (2017) STIM1 phosphorylation at Y361 recruits Orai1 to STIM1 puncta and induces Ca2+ entry. Sci Rep 7:42758. https://doi.org/10.1038/srep42758

Ying X, Minamiya Y, Fu C, Bhattacharya J (1996) Ca2+ waves in lung capillary endothelium. Circ Res 79:898–908

Zeng H, He X, Tuo QH, Liao DF, Zhang GQ, Chen JX (2016) LPS causes pericyte loss and microvascular dysfunction via disruption of Sirt3/angiopoietins/Tie-2 and HIF-2alpha/Notch3 pathways. Sci Rep 6:20931. https://doi.org/10.1038/srep20931

Zhou C, Townsley MI, Alexeyev M, Voelkel NF, Stevens T (2016) Endothelial hyperpermeability in severe pulmonary arterial hypertension: role of store-operated calcium entry. Am J Physiol Lung Cell Mol Physiol 311:L560–L569. https://doi.org/10.1152/ajplung.00057.2016

Chapter 4
Pulmonary Endothelial Cell Apoptosis in Emphysema and Acute Lung Injury

Eboni Chambers, Sharon Rounds, and Qing Lu

Abstract Apoptosis plays an essential role in homeostasis and pathogenesis of a variety of human diseases. Endothelial cells are exposed to various environmental and internal stress and endothelial apoptosis is a pathophysiological consequence of these stimuli. Pulmonary endothelial cell apoptosis initiates or contributes to progression of a number of lung diseases. This chapter will focus on the current understanding of the role of pulmonary endothelial cell apoptosis in the development of emphysema and acute lung injury (ALI) and the factors controlling pulmonary endothelial life and death.

List of Abbreviations

AAT	Alpha1-anti-trypsin
ADA	Adenosine deaminase
ADP	Adenosine-5′-diphosphate
APC	Activated protein C
ARDS	Acute respiratory distress syndrome
ARs	Adenosine receptors
ATF6	Transcription factor 6
ATGs	Autophagy-related genes
ATP	Adenosine-5′-triphosphate
BALF	Broncheoalveolar lavage fluid
CD39	ecto-5′-nucleotidase
CD73	ecto-5′-nucleotidase
CHOP	C/EBP homologous protein
COPD	Chronic obstructive pulmonary disease
CS	Cigarette smoke
CSE	Cigarette smoke extract

E. Chambers • S. Rounds • Q. Lu (✉)
Department of Medicine, Alpert Medical School of Brown University, Providence, RI, USA

Vascular Research Laboratory, Providence Veterans Affairs Medical Center, Providence, RI, USA
e-mail: Qing_Lu@brown.edu

K. Parthasarathi (ed.), *Molecular and Functional Insights Into the Pulmonary Vasculature*, Advances in Anatomy, Embryology and Cell Biology 228,
https://doi.org/10.1007/978-3-319-68483-3_4

DAMPs	Damage associated molecular patterns
EC	Endothelial cells
ECM	Extracellular matrix
eIF2α	Eukaryotic initiation factor 2α
ENT1/2	Equilibrative nucleoside transporter 1/2
ER	Endoplasmic reticulum
FAC	Focal adhesion complexes
FAK	Focal adhesion kinase
GSH	Glutathione
ICMT	Isoprenylcysteine-O-carboxyl methyltransferase
IRAK-1	Interleukin (IL)-1 receptor associated kinase
IRE1	Inositol-requiring enzyme 1
JNK	c-Jun N-terminal kinase
LPS	Lipopolysaccharide
MLKL	Mixed lineage kinase domain-like protein
mTOR	Mammalian target of rapamycin
MyD88	Myeloid differentiation factor 88
PERK	Pancreatic ER kinase like ER kinase
RBC	Red blood cells
RIPK1/3	Receptor-interacting protein kinase 1 and 3
ROS	Reactive oxygen species
S1P	Sphingosine 1-phosphate
SAH	S-adenosyl-L-homocysteine
SAHH	S-adenosyl-L-homocysteine hydrolase
SAM	S-Adenosyl-L-Methionine
TLRs	Toll-like receptors
TNF-α	Tumor necrosis factor-alpha
TRAF-6	TNF receptor associated factor-6
UPR	Unfolded protein response
VEGF	Vascular endothelial growth factor
VEGFR2	VEGF receptor type 2

4.1 Overview of Cell Death

4.1.1 Apoptosis

Apoptosis is a term first used by Kerr et al. in 1972 to describe a genetically determined energy-dependent active form of programmed cellular suicide. Apoptosis is characterized by well-ordered morphologic and molecular features including: cell surface exposure of phosphatidylserine, plasma membrane blebbing, cell shrinkage, cytoskeletal rearrangement, collapse of nuclear membrane, chromatin condensation, DNA fragmentation, and formation of membrane bound fragments known as "apoptotic bodies" (Kerr et al. 1972). Cell surface-exposed phosphatidylserine acts as a chemoattractant for phagocytes to engulf and clear apoptotic bodies (Henson and Tuder 2008). Apoptosis serves to eliminate unwanted, aged, harmful, injured, or infected cells. Due to limited release of intracellular contents, minimal inflammation occurs (Savill et al. 2002). However, if ingestion of apoptotic bodies by monocytes, macrophages, and dendritic cells (efferocytosis) is impaired, inflammation and autoimmunity may be enhanced (Gaipl et al. 2006). Apoptosis plays an essential role in the maintenance of tissue homeostasis and embryonic development. Further, during embryonic development, the timing of apoptosis is genetically determined. Excessive or inadequate apoptosis can, however, contribute to the pathogenesis of a variety of human diseases. Apoptosis is triggered by external stressors (e.g., death ligands, ultraviolet, and γ radiation) and/or internal stimuli (e.g., oxidants, DNA damage, increased Ca^{2+}). Apoptosis is processed by two fundamental signaling pathways: the death receptor-mediated extrinsic pathway and the mitochondria-dependent intrinsic pathway (Olson and Kornbluth 2001; Thorburn 2004). Extrinsic pathway-activated caspase-8 can truncate and activate BID, thus activating the intrinsic pathway (Li et al. 1998). The details on regulation of apoptosis have been reviewed (Harrington et al. 2007; Subramanian and Steer 2010; Ola et al. 2011). Therapies targeting regulators of apoptosis have been used in preclinical and clinical trials for a variety of diseases including the treatment of cancers (Goldar et al. 2015).

4.1.2 Necrosis

Necrosis is a passive and caspase-independent cell death, characterized by cell swelling, mitochondrial degeneration, impaired ATP generation, lysosomal leakage, early rupture of plasma membranes, random fragmentation/degradation of DNA, and leakage of cellular contents into the surrounding environment (Henriquez et al. 2008). Necrosis is usually induced by nonspecific and non-physiological stress. Further, inhibition of caspases leads to necrosis (Henriquez et al. 2008). Due to release of potentially pro-inflammatory and pro-immunogenic cellular contents into surrounding tissues, necrosis often induces inflammation, autoimmune responses, and is often seen concomitant with apoptosis.

4.1.3 Necroptosis

Necroptosis describes a type of active, regulated, and programmed necrosis dependent upon the serine/threonine kinase activity of receptor-interacting protein kinase 1 and 3 (RIPK1/3) (Linkermann and Green 2014). Necroptosis and apoptosis share several upstream signaling elements including death receptors caspase 8 and FLIP. When caspase-8 is inhibited, RIPK1 is activated and forms an intracellular complex with RIPK3 to assemble the necrosome, leading to phosphorylation of mixed lineage kinase domain-like protein (MLKL) and ultimately cell death. Unlike apoptosis, necroptosis promotes harmful innate and adaptive immunologic responses by releasing damage associated molecular patterns (DAMPs). Thus, the reduction of necroptosis might be beneficial by minimizing the release of DAMPs and proinflammatory responses. Necroptosis is, however, a defense mechanism against invading microbes, including viral infections, and promotes the death and removal of virally infected cells. Therefore, blockade of necroptosis may increase susceptibility to viral infections particularly in patients with suppressed immunity. A number of inhibitors of necroptosis, such as necrostatin (specific inhibitor for RIPK1) and necrosulfonamide (specific inhibitor for human MLKL), have been described, providing potential therapeutic tools for treatment. Given the complex role of necroptosis, tissue and cell-specific targeting therapy is needed.

4.1.4 Endoplasmic Reticulum Stress-Induced Apoptosis

The endoplasmic reticulum (ER) is the site of posttranslational modifications and folding of secreted and membrane proteins. A variety of insults, such as ER Ca^{2+} chelators, reducing agents, glucose starvation, glycosylation antagonists, and protein mutations, can disrupt ER protein folding and lead to an accumulation of unfolded or misfolded proteins in the ER, thus initiating ER stress (Schroder and Kaufman 2005). Cells respond to ER stress by the unfolded protein response (UPR). The UPR includes three arms: pancreatic ER kinase (PKR)-like ER kinase (PERK)/ eukaryotic initiation factor 2α (eIF2α), transcription factor 6 (ATF6), and inositol-requiring enzyme 1 (IRE1) (Schroder and Kaufman 2005). Through the UPR, cells attempt to restore ER homeostasis in order to maintain cell survival by inhibiting global protein synthesis (to reduce the loading of client protein to the ER for folding), enhancing ER protein folding capacity, and promoting ER-associated degradation of misfolded or unfolded proteins (Schroder and Kaufman 2005).

Prolonged ER stress causes cell death due to simultaneous activation of multiple apoptotic pathways by the UPR (Szegezdi et al. 2006). PERK-induced phosphorylation of eIF2α can lead to apoptosis by induction of pro-apoptotic transcription factor, C/EBP homologous protein (CHOP), which suppresses expression of anti-apoptotic protein, Bcl-2. Activated IRE1 activates c-Jun N-terminal kinase (JNK), which causes apoptosis by phosphorylation and thus inactivation of Bcl-2 and by phosphorylation and thus activation of pro-apoptotic protein, Bim. In addition, increased Ca^{2+} in the ER activates the death effector, Bax/Bak in the ER membrane, causing movement of Ca^{2+} from the ER to the mitochondria leading to mitochondrial-dependent apoptosis. ER membrane-localized caspase-12 (rodent) and caspase-4 (human) have also been implicated in ER-stress-induced apoptosis (Szegezdi et al. 2003; Kim et al. 2006). Caspase-12/-4 are cleaved and thus activated by the Ca^{2+}-dependent protease, m-calpain, by ER stress (Groenendyk and Michalak 2005). However, other studies have suggested that ER stress-induced apoptosis depends upon the apoptosome and not caspase-12/-4 (Obeng and Boise 2005; Di Sano et al. 2006).

Cell fate determination is not well understood when both survival (adaptive) and apoptotic pathways are simultaneously activated. It has been proposed that persistent ER stress causes apoptosis due to sustained induction of CHOP and instability of the adaptive pathway (Lin et al. 2007). It has also been suggested that cells survive mild ER stress because of the short half-life of pro-apoptotic proteins, compared to pro-survival proteins (Rutkowski et al. 2006). Robust prolonged ER stress causes apoptosis due to the induction of CHOP excessive to its degradation (Rutkowski et al. 2006).

4.1.5 Autophagy-Associated Cell Death

Autophagy is a dynamic and continuous process by which cells dispose of damaged or unneeded cellular proteins or organelles (mitochondria) by self-digestion to generate intracellular nutrients. During physiological conditions, autophagy is suppressed by mammalian target of rapamycin (mTOR), thus inhibiting the expression of autophagy-related genes (ATGs). Upon external or internal stress: including nutrient starvation, growth factor deprivation, hypoxia, ischemia, or mitochondrial aging, mTOR is inhibited thus initiating autophagy. Autophagy is a multistep sequential process, consisting of the formation of double-membrane vesicles that sequester unwanted cargo (proteins or mitochondria) in autophagosomes, fusion of autophagosomes with endosomes or lysosomes to form amphisomes or autolysosomes, and digestion of cargo by proteases (Hotchkiss et al. 2009; Choi et al. 2013). Autophagy is an evolutionarily conserved housekeeping process that allows recycling of damaged proteins and organelles in order to maintain homeostasis. Impairment in any step of autophagy causes cellular nutrient deficiency and/or accumulation of damaged proteins and organelles leading to cell death (Hotchkiss et al. 2009). Whether autophagy promotes cell survival or death may depend on cell type and setting (Gustafsson and Gottlieb 2008).

4.1.6 Assessments of Cell Death

Based on the unique characteristics of different types of cell death, a variety of assays have been developed to assess the specific types of cell death in vivo and in vitro. Different types of cell death may share common characteristics at different stages of cell death; therefore, it is often necessary to use multiple assays to confirm cell death. The details on the assessments of cell death have been extensively reviewed (Harrington et al. 2007; Henson and Tuder 2008; Lu and Rounds 2009; Klionsky et al. 2016) and will not be discussed in this review.

4.2 Pulmonary Endothelial Cell Apoptosis

Balance of endothelial cell survival and death is crucial for angiogenesis, vessel regression, and barrier function. Due to the unique position of endothelial cells (EC) at the interface of circulating blood and surrounding tissues, EC may be exposed to various environmental stress including: hypoxia, hyperoxia, oxidants, lipopolysaccharide (LPS), and cigarette smoke (CS), or internal stress including: adenosine, ceramide, tumor necrosis factor (TNF)-α, and angiotensin II. Apoptosis is a pathophysiological consequence of these stimuli. However, a variety of biomechanical and biochemical factors are involved in the anti-apoptotic processes. For example, physiological levels of shear stress and cyclic strain, vascular endothelial growth factor (VEGF), focal adhesion kinase (FAK), activated protein C (APC), and sphingosine 1-phosphate (S1P) protect EC against apoptosis. The pro- and anti-apoptotic effects of these mediators have been reviewed (Harrington et al. 2007; Lu and Rounds 2009); therefore, this review will focus on the current understanding of endothelial pro-survival factors (VEGF and FAK) and apoptosis-inducing stress (adenosine, cigarette smoke, and LPS) in the lungs.

4.2.1 Vascular Endothelial Growth Factor

EC express abundant VEGF, which promotes EC survival and maintains normal alveolar structure (Voelkel et al. 2006). Expression of both VEGF and VEGF receptor type 2 (VEGFR2) are decreased in lung tissue of patients with chronic obstructive pulmonary disease (COPD) (Kasahara et al. 2001). This diminished VEGF/VEGFR2 signaling is inversely associated with increased lung EC apoptosis (Kasahara et al. 2001). Lung-targeted inhibition of VEGF or VEGFR2 causes alveolar septal cell apoptosis in mice (Kasahara et al. 2000; Tang et al. 2004). Our group has also shown that blockade of VEGFR2 causes cultured pulmonary artery EC apoptosis in vitro (Lu 2008). These results indicate that VEGF signaling is essential for lung EC survival.

4.2.2 Focal Adhesion Kinase

EC are linked to the basement membrane through binding of cell surface expressed integrins to extracellular matrix (ECM) proteins at focal adhesion complexes (FAC) (Hynes 1992). As anchorage-dependent cells, EC undergo detachment-initiated apoptosis, referred to as anoikis, upon loss of adhesion to underlying basement membrane. FAK, a non-receptor tyrosine kinase and an essential component of FAC, is activated upon integrin engagement of ECM (Guan et al. 1991; Guan and Shalloway 1992; Parsons 2003). FAK provides survival signaling for anchorage-dependent cells such as cultured fibroblasts (Hungerford et al. 1996). Similarly, EC isolated from FAK-null embryos undergo apoptosis (Ilic et al. 1995, 2003). Endothelium-specific deletion of FAK (Cre/FAK^{flox}) is embryonic lethal and causes EC apoptosis (Shen et al. 2005; Braren et al. 2006). Guan and colleagues (Guan et al. 1991; Guan and Shalloway 1992) have demonstrated that FAK tyrosine kinase activity is essential for FAK activity. FAK promotes cell survival by recruiting proteins containing SH2 domain including Src and phosphatidylinositol-3-kinase (PI3K) (Schaller et al. 1994). The activated PI3K recruits and activates Akt (Khwaja et al. 1997), which promotes cell survival via phosphorylation and thus inhibition of pro-apoptotic protein, Bad (Kennedy et al. 1997). FAK also promotes survival by activation of NF-κB and ERK signaling pathways (Huang et al. 2007). Additionally, FAK can translocate to the nucleus and inhibit p53 transcriptional activation and enhance p53 degradation, leading to protection against apoptosis (Ilic et al. 1998).

4.2.3 Adenosine

Adenosine is generated from adenosine-5′-triphosphate (ATP) and adenosine-5′-diphosphate (ADP) by extracellular ecto-5′-nucleotidases, CD39 and CD73, and is metabolized by adenosine deaminase (ADA). Extracellular adenosine exists in low concentrations (40–600 nM) under physiological conditions and is increased due to platelet degranulation, cell necrosis, activation of CD39 and/or CD73, or inhibition of ADA (Thompson et al. 2004; Eltzschig et al. 2006; Volmer et al. 2006; Eckle et al. 2007). Increased extracellular adenosine can interact with cell surface G-protein coupled adenosine receptors (ARs) (Feoktistov et al. 2002; Wyatt et al. 2002; Umapathy et al. 2010). Activation of adenosine receptors, specifically A_3-mediated signaling, has been shown to protect against apoptosis and tissue injury (Rivo et al. 2004; Chen et al. 2006; Matot et al. 2006).

However, sustained increased adenosine in ADA-deficient mice enhances alveolar cell apoptosis (Zhou et al. 2009). We have also shown that prolonged exposure to adenosine causes apoptosis of cultured lung EC (Lu et al. 2013). The injurious effect of adenosine is mediated by equilibriative nucleoside transporters. EC predominantly express equilibriative nucleoside transporter 1 (ENT_1) and ENT_2 (Archer et al. 2004). Upon sustained exposure, adenosine may be taken up into cells by ENTs. Further, similar to other G-protein coupled receptors, prolonged engagement of ARs causes receptor desensitization and internalization (Fredholm et al. 2001). This concept is supported by findings that sustained increased adenosine in ADA-deficient mice enhances alveolar cell apoptosis via a mechanism independent of adenosine receptor, $A_{2B}R$ (Zhou et al. 2009). In addition, sustained exposure to adenosine causes endothelial cell apoptosis; this effect is prevented by inhibition of $ENT_{1/2}$ however exacerbated by inhibition of either $A_{2A}R$ or $A_{2B}R$ (Lu et al. 2013). These results are consistent with the concept that $ENT_{1/2}$-facilitated intracellular adenosine uptake and subsequent metabolism mediates adenosine-induced EC apoptosis, whereas AR-mediated signaling limits apoptosis (Simonis et al. 2009).

Once intracellular, adenosine reacts with homocysteine and generates S-adenosyl-L-homocysteine (SAH) by inhibition of SAH hydrolase (SAHH). SAH induces endothelial cell apoptosis independent of homocysteine (Sipkens et al. 2012). SAH is also a product of carboxyl methylation with S-adenosyl-L-methionine (SAM) as a methyl donor. We have demonstrated that exogenous adenosine causes lung EC apoptosis via increased ratio of intracellular SAH to SAM (Rounds et al. 1998). The increased ratio of SAH to SAM suppresses carboxyl methyltransferase activity. Isoprenylcysteine-O-carboxyl methyltransferase (ICMT) is a major methyltransferase for carboxyl methylation of small GTPase, Ras (Clarke 1992), which is a posttranslational modification essential for membrane localization and activation of Ras (Boivin and Beliveau 1995; Fleming et al. 1996; Kranenburg et al. 1997; Michaelson et al. 2001). We have shown that exogenous adenosine causes lung EC apoptosis in part by ICMT inhibition-mediated inhibition of Ras carboxyl methylation and activation (Kramer et al. 2003).

SAM is a precursor to glutathione (GSH) and is synthesized exclusively in the cytosol (Reytor et al. 2009) and also transported into mitochondria (Agrimi et al. 2004). Exogenous SAM has been shown to elevate GSH levels in vivo and prevent alcohol-induced mitochondrial oxidative stress and dysfunction as well as liver and lung injury in animal models (Holguin et al. 1998; Bailey et al. 2006; Cederbaum 2011). p38 is a redox-sensitive protein (Matsuzawa and Ichijo 2008). Reactive oxygen species (ROS)-mediated p38 activation has been implicated in extracellular ATP-induced macrophage apoptosis (Noguchi et al. 2008) and H_2O_2-induced EC apoptosis (Machino et al. 2003). Activation of p38 has also been implicated in homocysteine-induced apoptosis of endothelial progenitor cells (Bao et al. 2010) and cardiomyocytes (Wang et al. 2011). We have shown that sustained exposure to exogenous adenosine causes mitochondrial defects and endothelial apoptosis via mitochondrial oxidative stress-induced activation of p38 (Lu et al. 2012, 2013). Active p38 causes apoptosis by direct phosphorylation, and thus inhibition of Bcl-2 (De Chiara et al. 2006; Farley et al. 2006) and by increasing mitochondrial translocation of Bax (Capano and Crompton 2006). Future studies are needed to address whether sustained adenosine exposure reduces mitochondrial SAM, thus leading to mitochondrial oxidative stress via increased ratio of SAH to SAM in the cytosol.

In summary, adenosine displays seemingly paradoxical effects on lung EC life and death. Acute exposure protects EC against apoptosis via AR-mediated signaling, whereas prolonged exposure causes EC apoptosis via $ENT_{1/2}$-mediated intracellular adenosine uptake and subsequent metabolism and mitochondrial oxidative stress.

4.2.4 Cigarette Smoke

Lung EC apoptosis is significantly elevated in human smokers with emphysema (Kasahara et al. 2001) and mice with mild emphysema caused by CS exposure (Sakhatskyy et al. 2014). We (Sakhatskyy et al. 2014) and others (Tuder et al. 2000; Damico et al. 2011) have shown that CS extract (CSE) causes cultured lung macro- and microvascular EC apoptosis in vitro. The mechanisms underlying CS-induced lung EC apoptosis are rather complicated and involve FAK, p53, UPR, and autophagy.

FAK is a survival signal for anchorage-dependent cells (Hungerford et al. 1996). Tyrosine 397 phosphorylation of FAK is essential for its activation (Schaller et al. 1994). CSE decreases FAK phosphorylation at tyrosine-397 in an oxidative stress-dependent manner (Lu et al. 2011)—essential in CSE-induced EC apoptosis (Sakhatskyy et al. 2014). FAK also promotes cell survival via suppression of p53 (Ilic et al. 1998). Further, activation of p53 has contributed to CSE-induced pulmonary EC apoptosis (Damico et al. 2011). Thus, we speculate that CSE causes lung EC apoptosis via oxidative stress-mediated inhibition of FAK and subsequent activation of p53.

The UPR is an important mechanism of the elimination of ER stress and enhanced cell survival (Schroder and Kaufman 2005). The UPR is activated in lung tissue of smokers who do not have emphysema (Kelsen et al. 2008). The UPR is also activated by CSE in cultured human bronchial epithelial cells and 3T3 fibroblasts (Hengstermann and Müller 2008; Jorgensen et al. 2008) and cultured pulmonary EC (Sakhatskyy et al. 2014). Using mouse models of CS exposure, we have demonstrated a strong link between impairment of eIF2α signaling with lung EC apoptosis (Sakhatskyy et al. 2014). Future studies are necessary to determine if impaired eIF2α signaling contributes to lung EC apoptosis.

Autophagy is increased in response to deficiencies in extracellular and intracellular nutrients. Enhanced autophagy is observed in the lung tissue of smokers with emphysema (Chen et al. 2008). Autophagy is also activated by CSE exposure in lung epithelial cells and fibroblasts (Kim et al. 2008) as well as lung EC (Sakhatskyy et al. 2014). Increased autophagy has contributed to CS-induced alveolar epithelial cell apoptosis in mice (Chen et al. 2010). In contrast, increased autophagy has also been shown to protect against pulmonary endothelial cell apoptosis induced by cadmium, a component of cigarette smoke (Surolia et al. 2015). We have reported that autophagy was not altered in the lung tissue of a mouse strain susceptible to CS-induced lung EC apoptosis and emphysema (Sakhatskyy et al. 2014). The role of autophagy in CS-induced apoptosis may be dependent on cell types and stimuli.

Due to open structure and limited repair capacity, mitochondrial DNA is 50 times more sensitive to oxidative damage than nuclear DNA (Yakes and Van Houten 1997). Oxidative stress-induced mitochondrial DNA damage triggers mitochondrial dysfunction and apoptosis of lung EC (Ruchko et al. 2005). The role of mitochondrial DNA damage in CS-induced lung EC apoptosis remains to be studied.

4.2.5 Lipopolysaccharide

LPS, also known as lipoglycans or endotoxin, is a component of the outer envelope of gram-negative bacteria and elicits pro-inflammatory responses. It is well established that LPS-induced EC activation, dysfunction, and apoptosis play an important role in bacterial sepsis and endotoxemia. In the blood circulation, LPS binds to soluble CD14 via LPS-binding protein (LBP), followed by engagement of toll-like receptor

(TLR)-4. This engagement results in the recruitment of adaptor, myeloid differentiation factor 88 (MyD88), and subsequent activation of interleukin (IL)-1 receptor associated kinase (IRAK)-1, TNF receptor associated (TRAF)-6, NF-kB, and MAPK pathways (Desch et al. 1989; Wang et al. 2001; Bannerman and Goldblum 2003).

NF-kB has been shown to transcriptionally upregulate anti-apoptotic genes such as IAP-1, IAP-2, and FLIP (LaCasse et al. 1998; Bannerman et al. 2004). However, suppression of NF-kB has minimal effect on LPS-induced EC apoptosis (Zen et al. 1999). This is due to FADD/MyD88-dependent negative regulation of LPS-induced NF-kB activation (Martin et al. 2005; Zhande et al. 2007); Fas is no longer able to activate MyD88, thus stimulating LPS/TLR4/NF-kB signaling (Martin et al. 2005). LPS also stimulates MyD88-independent signaling of endothelial apoptosis (Dauphinee and Karsan 2006). Heterotrimeric Gi/Go proteins play a role in LPS-induced TLR signaling independent of the MyD88-dependent pathway, leading to MAPK, Akt, and IFN activation of endothelial cells (Dauphinee et al. 2011). Whether LPS-induced stimulation of heterotrimeric G coupled proteins plays a role in EC apoptosis is unknown. LPS can activate the BID-dependent intrinsic pathway of apoptosis in lung EC (Wang et al. 2007). Conversely, LPS has been shown to upregulate mRNA of anti-apoptotic molecules, thus preventing EC apoptosis (Hu et al. 1998). LPS-induced intrinsic apoptosis and cytoprotection in disease states are not well understood and require further study.

4.3 Pulmonary EC Apoptosis in Lung Diseases

Apoptosis has been shown to ameliorate or exacerbate lung injury. Pulmonary EC apoptosis plays an important role in physiological processes including vasculogenesis and angiogenesis during lung development. Pulmonary EC apoptosis may also initiate or contribute to the progression of a number of lung diseases, as reviewed elsewhere (Harrington et al. 2007; Lu and Rounds 2009). In this review, we will focus on the role of pulmonary EC apoptosis in development of emphysema and Acute Lung Injury (ALI).

4.3.1 Emphysema

Chronic obstructive pulmonary disease (COPD), a progressive respiratory condition consisting of emphysema and chronic bronchitis, is the fourth leading cause of death worldwide and may become the third leading cause of death by 2030 based on prediction by the World Health Organization (Khaltaev 2005). The prevalence of COPD in the United States in 2013 was estimated to be 6.4% (15.7 million adults) (Wheaton et al. 2015). COPD is also an important contributor of mortality and disability in the United States (Murray et al. 2013). Further, COPD-related medical costs were estimated at $32 billion in the USA in 2010 with an additional $4 billion in costs due to absence from work (Ford et al. 2015). α1-antitrypsin (AAT) deficiency and other genetic predispositions contribute to the development of COPD (Sandford et al. 1997). However, tobacco smoke remains the leading cause of this devastating disease. Indoor air pollution (such as biomass fuel used for cooking and heating), outdoor air pollution, and occupational dusts and chemicals also increase the risk of COPD (Diette et al. 2012). Although the pathology of COPD has been well defined, the pathogenesis of the disease initiation and progression is not understood. Currently, there is no specific treatment available to reverse COPD.

Emphysema, a common and debilitating manifestation of COPD, is characterized by alveolar airspace enlargement, loss of alveolar capillary septa, and resultant impaired gas exchange. Several hypotheses have been proposed to explain alveolar wall damage in emphysema. Protease/anti-protease imbalance has been accepted as a major mechanism for emphysematous lung destruction (Shapiro 1995, 1999; Shapiro et al. 2003; Taraseviciene-Stewart and Voelkel 2008). It is believed that neutrophil elastase and macrophage matrix metalloproteinases enzymatically degrade elastin in alveolar septa, leading to emphysema (Taraseviciene-Stewart and Voelkel 2008). This notion is supported by findings that patients with genetic deficiency of the anti-protease, AAT, develop emphysema (No Authors 1997). Additionally, intra-tracheal instillation of proteases causes an emphysema phenotype in rats (Pastor et al. 2006). However, less than 5% of emphysema patients have AAT deficiency. Inflammatory cell infiltration is also seen in human emphysema. However, lung inflammation in pneumonia or acute lung injury does not usually result in emphysema. This suggests that inflammation may not be sufficient by itself for the development of emphysema. Oxidant stress and immunological injury also play a role in the pathogenesis of emphysema (Taraseviciene-Stewart and Voelkel 2008). Emerging evidence has highlighted a role of apoptosis, particularly EC apoptosis, in the initiation and progression of emphysema (Kasahara et al. 2000, 2001; Giordano et al. 2008).

Lung tissue from patients with emphysema displays increased apoptosis of both epithelial and endothelial cells in the alveolar septa (Kasahara et al. 2001; Imai et al. 2005). Bcl-2 single-nucleotide polymorphisms have been associated with severity of human emphysema (Sata et al. 2007). We have shown that lung EC apoptosis is elevated in a mouse model of emphysema induced by CS exposure (Sakhatskyy et al. 2014). Interestingly, induction of alveolar cell apoptosis by intratracheal instillation of the active caspase-3 causes emphysema in rats (Aoshiba et al. 2003). Additionally, inhibition of VEGF signaling causes alveolar septal cell apoptosis and emphysema in mice (Kasahara et al. 2000; Tang et al. 2004). Similarly, intra-tracheal instillation of C_{12} ceramide triggers alveolar endothelial and epithelial cell apoptosis and emphysema-like changes in mice (Petrache et al. 2005). Further, lung EC-targeted induction of apoptosis led to emphysema and enhanced oxidative stress and lung inflammation (Giordano et al. 2008). More importantly, inhibition of apoptosis using pan-caspase inhibitors prevented the emphysematous changes induced by either ceramide (Petrache et al. 2005) or blockage of VEGF signaling (Kasahara et al. 2000; Tang et al. 2004). These results support a central role of lung EC apoptosis in the development of emphysema. Anti-protease, AAT, inhibits CSE-induced pulmonary EC apoptosis in vitro by direct interaction with caspase-3 (Aldonyte et al. 2008). Overexpression of AAT also inhibits lung endothelial apoptosis and attenuates emphysema caused by either active caspase-3 or blockade of VEGF signaling (Petrache et al. 2006). These studies suggest that lung EC apoptosis is a critical step in the pathogenesis of emphysema.

Inhibition of FAK causes emphysema-like change in rat lungs (Mizuno et al. 2012). We have shown that CS exposure for 3 weeks enhanced pulmonary EC apoptosis and decreased FAK activity in mice susceptible to CS-induced emphysema (Sakhatskyy et al. 2014). Further studies are necessary to address whether reduced FAK activity contributes to CS-induced lung EC apoptosis and emphysema in humans in vivo. We have shown that CS exposure increases lung tissue adenosine levels in mice, an effect associated with lung EC apoptosis and early emphysema (Lu et al. 2013). Sustained increased adenosine in ADA-deficient mice also enhances alveolar cell apoptosis and causes emphysema in mice (Zhou et al. 2009). ADA expression and activity are reduced in the lung of smokers with COPD (Zhou et al. 2010). Whether chronically elevated adenosine contributes to CS-induced lung endothelial cell apoptosis and development of emphysema remains to be investigated.

Ceramide is upregulated in emphysematous lungs of patients and animal models, as well as in cultured pulmonary EC exposed to CSE (Petrache et al. 2005). This increase in ceramide is associated with enhanced alveolar cell apoptosis (Petrache et al. 2005). Interestingly, intratracheal instillation of C_{12} ceramide triggers air-space enlargement and apoptosis of alveolar EC and type II epithelial cells (Petrache et al. 2005). Further, inhibition of de novo ceramide synthesis significantly attenuated lung cell apoptosis and emphysema induced by VEGFR2 blockade (Petrache et al. 2005). These results suggest that ceramide is also an important mediator of alveolar cell apoptosis and emphysema (Petrache et al. 2005).

Only 10–15% of smokers develop emphysema. The mechanism underlying increased susceptibility to emphysema remains unclear. The UPR is elevated in the lungs of smokers without evidence of emphysema (Kelsen et al. 2008). Nrf2, a redox-sensitive, antioxidant transcription factor, is activated by eIF2α, a branch of UPR (Digaleh et al. 2013). Nrf2 knockout mice demonstrate enhanced susceptibility to cigarette smoke-induced emphysema in comparison to wild-type mice (Iizuka et al. 2005). We have shown that active eIF2α was significantly reduced in the lungs of AKR mice with mild emphysema induced by CS (Sakhatskyy et al. 2014). Future studies are needed to address whether Nrf2 is reduced in the lungs and whether inadequate induction of Nrf2 contributes to development of emphysema.

Autophagy is significantly increased in lung tissue of patients with COPD; the degree of autophagy positively correlates with the clinical severity of disease (Chen et al. 2008). Increased autophagy has contributed to CS-induced alveolar epithelial cell apoptosis and emphysema in mice (Chen et al. 2010; Mizumura et al. 2014). In contrast, increased autophagy protects against pulmonary endothelial cell apoptosis and emphysema induced by cadmium, a component of cigarette smoke

(Surolia et al. 2015). We have reported that autophagy was not altered in lung tissue of a mouse strain with increased lung EC apoptosis and mild emphysema induced by CS (Sakhatskyy et al. 2014). Thus, the role of autophagy in regulating lung EC apoptosis and early onset of CS-induced emphysema needs further study.

4.3.2 Acute Lung Injury

ALI and its more severe form, acute respiratory distress syndrome (ARDS), are life-threatening disorders clinically characterized by severe hypoxemia and pulmonary bilateral infiltrates. In the United States, ARDS affects approximately 190,000 patients annually (Rubenfeld et al. 2005). ARDS accounts for 3.6 million associated hospital days (Rubenfeld et al. 2005; Adhikari et al. 2010). The global impact of ARDS has been difficult to assess due to varying definitions of the broad clinical phenotypes and limited data. Thus, ARDS remains an underreported disease of treated incidence, as opposed to actual incidence, in the undeveloped world (Buregeya et al. 2014). Although the mortality rate of ARDS has decreased to around 30–40% due to lung protective ventilation strategies (Amato et al. 1998; Villar et al. 2006), ARDS remains a deadly syndrome without a specific cure. Currently, there are no pharmacological interventions available to reduce the mortality of ARDS.

Sepsis, bacterial and viral pneumonia, and trauma remain the leading risk factors for the development of ARDS. Emerging evidence from epidemiologic studies, animal models, and cultured cell models have suggested that both active and passive cigarette smoke exposure modifies the susceptibility for development of ALI and ARDS (Iribarren et al. 2000; Calfee et al. 2011; Lu et al. 2011, 2013; Hsieh et al. 2014; Borgas et al. 2016).

The pathophysiology of ARDS is characterized by increased permeability of the alveolar-capillary barrier, influx of protein and inflammatory cell-rich fluid into the alveolar space, attenuated gas exchange between alveolar-capillary barrier, and dysregulated inflammation. Increased permeability of the microvascular endothelium and alveolar epithelium promotes edema formation, and this concept has been accepted as an important mechanism for the initiation of ARDS (Matthay et al. 2012). It is well established that polymorphonuclear cells (PMN) and immunological injury also play a significant role in the pathogenesis of ARDS (Perl et al. 2011). PMN accumulation is observed in the broncheoalveolar lavage fluid (BALF) (Pittet et al. 1997) and lung biopsies of early ARDS patients (Bachofen and Weibel 1977, 1982). Further, neutrophilia has been correlated with exacerbation of sepsis-induced ALI (Steinberg et al. 1994). However, ARDS may also develop in neutropenic patients, and neutrophil activation and migration may be observed in human lungs without injury (Martin et al. 1989; Downey et al. 1999). This suggests that inflammation may not be sufficient by itself for the development of ARDS.

Emerging evidence has suggested a role of pulmonary cell apoptosis in the initiation and progression of ARDS. The death receptor, Fas, and its ligand, FasL system, is an important death receptor-mediated extrinsic pathway of apoptosis. FasL is expressed and released by inflammatory cells, including neutrophils and lymphocytes, whereas Fas is expressed on the surface of lung EC, alveolar and bronchial epithelial cells, Clara cells, and alveolar macrophages. Fas and FasL are increased in pulmonary edema fluid and in lung tissue of patients with ARDS (Albertine et al. 2002). Silencing of Fas/FasL reduces lung cell apoptosis and ALI in a mouse model of sepsis (Perl et al. 2005, 2007). Soluble FasL (sFasL) is a cleaved form of FasL by metalloproteinases and is increased in BAL fluid of patients with ARDS (Matute-Bello et al. 1999). sFasL released from inflammatory cells is capable of inducing lung epithelial cell apoptosis (Matute-Bello et al. 1999). The role of Fas/FasL in lung EC apoptosis is not yet clear. Robust pulmonary endothelial cell apoptosis has been observed in patients with severe ARDS (Abadie et al. 2005) and in mice with ALI induced by LPS (Fujita et al. 1998). Sepsis-induced ARDS in mice indicates evidence for pulmonary microvascular endothelial cell death as a cause of barrier dysfunction and edema (Gill et al. 2014, 2015). Inhibition of apoptosis using a broad-spectrum caspase inhibitor prolonged survival of mice exposed to LPS (Kawasaki et al. 2000). Since apoptosis of alveolar endothelial, epithelial, and interstitial inflammatory cells occurs during ALI, future studies are needed to address the role of apoptosis of specific cells in initiation of ALI/ARDS.

Apoptosis has been thought of to be a non-inflammatory means of removing injurious cells, thus facilitating lung repair. However, there is increasing evidence indicating that Fas/FasL-mediated lung epithelial apoptosis results in release of pro-inflammatory cytokines (such as TNF-α and TGF-β1), leading to inflammation and progression from ARDS to fibrosis (Chapman 1999). Whether pulmonary endothelial cell apoptosis occurs during initiation or progression of pulmonary fibrosis is unknown.

The role of necroptosis in development of ARDS is yet to be determined. Of interest, a recent study of blood transfusion-related acute lung injury indicates that banked red blood cell (RBC) transfusion enhances susceptibility to lung inflammation and ARDS in critically ill transfused patients and mice through necroptosis of lung EC and subsequent release of DAMPs (Qing et al. 2014).

4.4 Conclusions and Perspectives

Cell life and death are tightly regulated by survival signaling and death inducing programs. Pulmonary EC apoptosis significantly contributes to the development of emphysema and ALI/ARDS, as depicted in Fig. 4.1. Pan-caspase inhibitors have been used to inhibit lung cell apoptosis and prevent emphysema and ALI in animal models. However, use of such drugs to treat apoptosis-associated lung diseases may be problematic due to breakdown of tissue homeostasis and activation of necroptosis (Linkermann and Green 2014). The therapeutic potential of drugs that modulate cell death is dependent upon cell type-specific, tissue-specific, and

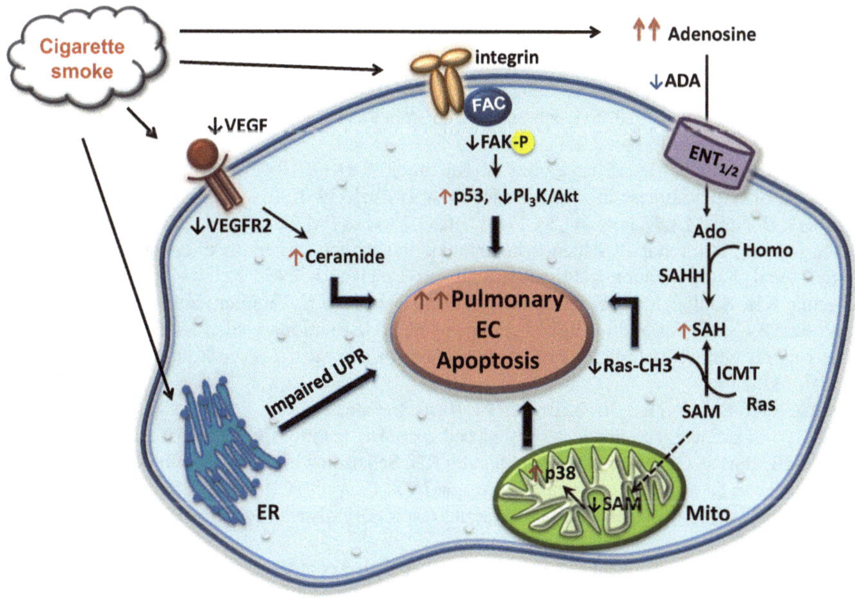

Fig. 4.1 Signaling pathways to CS-induced pulmonary endothelial cell apoptosis. Multiple signaling pathways are involved in CS-induced pulmonary endothelial cell apoptosis. (1) CS reduces VEGF/VEGFR2 signaling, leading to induction of ceramide and consequent apoptosis; (2) CS reduces FAK activation, leading to activation of p53 and inhibition of PI3K/Akt signaling, which results in apoptosis; (3) CS causes mitochondrial oxidative stress and mitochondrial dysfunction, leading to apoptosis; (4) CS elevates adenosine levels, leading to inactivation of Ras and mitochondrial oxidative stress, resulting in apoptosis; (5) CS impairs unfolded protein response, leading to apoptosis

vascular bed-specific actions. Thus, drugs acting locally and with cell type specificity are needed. Areas where research is needed include: (1) apoptosis susceptibility of different EC (conduit artery versus microvascular versus progenitor); (2) role of apoptosis of specific lung cells in initiation and/or progression of lung diseases; (3) role of necrosis and necroptosis in development of lung diseases, such as emphysema and ALI.

Acknowledgments This work was supported with the use of facilities at the Providence VA Medical Center and by an Institutional Development Award (IDeA) from the National Institute of General Medical Sciences of the National Institutes of Health under grant number P20GM103652 (Lu, project 1), P20GM103652 (Rounds), VA Merit Review (Rounds), NIH Ro1 HL130230 (Lu).

References

Abadie Y, Bregeon F, Papazian L, Lange F, Chailley-Heu B, Thomas P, Duvaldestin P, Adnot S, Maitre B, Delclaux C (2005) Decreased VEGF concentration in lung tissue and vascular injury during ARDS. Eur Respir J 25(1):139–146

Adhikari NK, Fowler RA, Bhagwanjee S, Rubenfeld GD (2010) Critical care and the global burden of critical illness in adults. Lancet 376(9749):1339–1346

Agrimi G, Di Noia MA, Marobbio CM, Fiermonte G, Lasorsa FM, Palmieri F (2004) Identification of the human mitochondrial S-adenosylmethionine transporter: bacterial expression, reconstitution, functional characterization and tissue distribution. Biochem J 379(Pt 1):183–190

Albertine KH, Soulier MF, Wang Z, Ishizaka A, Hashimoto S, Zimmerman GA, Matthay MA, Ware LB (2002) Fas and fas ligand are up-regulated in pulmonary edema fluid and lung tissue of patients with acute lung injury and the acute respiratory distress syndrome. Am J Pathol 161 (5):1783–1796

Aldonyte R, Hutchinson ET, Jin B, Brantly M, Block E, Patel J, Zhang J (2008) Endothelial alpha-1-antitrypsin attenuates cigarette smoke induced apoptosis in vitro. COPD 5:153–162

Amato MB, Barbas CS, Medeiros DM, Magaldi RB, Schettino GP, Lorenzi-Filho G, Kairalla RA, Deheinzelin D, Munoz C, Oliveira R, Takagaki TY, Carvalho CR (1998) Effect of a protective-ventilation strategy on mortality in the acute respiratory distress syndrome. N Engl J Med 338 (6):347–354

Aoshiba K, Yokohori N, Nagai A (2003) Alveolar wall apoptosis causes lung destruction and emphysematous changes. Am J Respir Cell Mol Biol 28:555–562

Archer RG, Pitelka V, Hammond JR (2004) Nucleoside transporter subtype expression and function in rat skeletal muscle microvascular endothelial cells. Br J Pharmacol 143(1):202–214

Bachofen M, Weibel ER (1977) Alterations of the gas exchange apparatus in adult respiratory insufficiency associated with septicemia. Am Rev Respir Dis 116(4):589–615

Bachofen M, Weibel ER (1982) Structural alterations of lung parenchyma in the adult respiratory distress syndrome. Clin Chest Med 3(1):35–56

Bailey SM, Robinson G, Pinner A, Chamlee L, Ulasova E, Pompilius M, Page GP, Chhieng D, Jhala N, Landar A, Kharbanda KK, Ballinger S, Darley-Usmar V (2006) S-adenosylmethionine prevents chronic alcohol-induced mitochondrial dysfunction in the rat liver. Am J Physiol Gastrointest Liver Physiol 291(5):G857–G867

Bannerman DD, Goldblum SE (2003) Mechanisms of bacterial lipopolysaccharide-induced endothelial apoptosis. Am J Physiol Lung Cell Mol Physiol 284(6):L899–L914

Bannerman DD, Eiting KT, Winn RK, Harlan JM (2004) FLICE-like inhibitory protein (FLIP) protects against apoptosis and suppresses NF-kappaB activation induced by bacterial lipopolysaccharide. Am J Pathol 165(4):1423–1431

Bao XM, Wu CF, Lu GP (2010) Atorvastatin inhibits homocysteine-induced oxidative stress and apoptosis in endothelial progenitor cells involving Nox4 and p38MAPK. Atherosclerosis 210 (1):114–121

Boivin D, Beliveau R (1995) Subcellular distribution and membrane association of Rho-related small GTP-binding proteins in kidney cortex. Am J Phys 269:F180–F189

Borgas D, Chambers E, Newton J, Ko J, Rivera S, Rounds S, Lu Q (2016) Cigarette smoke disrupted lung endothelial barrier integrity and increased susceptibility to acute lung injury via histone deacetylase 6. Am J Respir Cell Mol Biol 54(5):683–696

Braren R, Hu H, Kim YH, Beggs HE, Reichardt LF, Wang R (2006) Endothelial FAK is essential for vascular network stability, cell survival, and lamellipodial formation. J Cell Biol 172 (1):151–162

Buregeya E, Fowler RA, Talmor DS, Twagirumugabe T, Kiviri W, Riviello ED (2014) Acute respiratory distress syndrome in the global context. Glob Heart 9(3):289–295

Calfee CS, Matthay MA, Eisner MD, Benowitz N, Call M, Pittet JF, Cohen MJ (2011) Active and passive cigarette smoking and acute lung injury after severe blunt trauma. Am J Respir Crit Care Med 183(12):1660–1665

Capano M, Crompton M (2006) Bax translocates to mitochondria of heart cells during simulated ischaemia: involvement of AMP-activated and p38 mitogen-activated protein kinases. Biochem J 395(1):57–64

Cederbaum AI (2011) Hepatoprotective effects of S-adenosyl-L-methionine against alcohol- and cytochrome P450 2E1-induced liver injury. World J Gastroenterol 16(11):1366–1376

Chapman HA (1999) A Fas pathway to pulmonary fibrosis. J Clin Invest 104:1–2

Chen GJ, Harvey BK, Shen H, Chou J, Victor A, Wang Y (2006) Activation of adenosine A3 receptors reduces ischemic brain injury in rodents. J Neurosci Res 84(8):1848–1855

Chen ZH, Kim HP, Sciurba FC, Lee SJ, Feghali-Bostwick C, Stolz DB, Dhir R, Landreneau RJ, Schuchert MJ, Yousem SA, Nakahira K, Pilewski JM, Lee JS, Zhang Y, Ryter SW, Choi AM (2008) Egr-1 regulates autophagy in cigarette smoke-induced chronic obstructive pulmonary disease. PLoS One 3(10):e3316

Chen ZH, Lam HC, Jin Y, Kim HP, Cao J, Lee SJ, Ifedigbo E, Parameswaran H, Ryter SW, Choi AM (2010) Autophagy protein microtubule-associated protein 1 light chain-3B (LC3B) activates extrinsic apoptosis during cigarette smoke-induced emphysema. Proc Natl Acad Sci U S A 107(44):18880–18885

Choi AM, Ryter SW, Levine B (2013) Autophagy in human health and disease. N Engl J Med 368 (7):651–662

Clarke S (1992) Protein isoprenylation and methylation at carboxyl terminal cysteine residues. Annu Rev Biochem 61:355–386

Damico R, Simms T, Kim BS, Tekeste Z, Amankwan H, Damarla M, Hassoun PM (2011) p53 mediates cigarette smoke-induced apoptosis of pulmonary endothelial cells: inhibitory effects of macrophage migration inhibitor factor. Am J Respir Cell Mol Biol 44(3):323–332

Dauphinee SM, Karsan A (2006) Lipopolysaccharide signaling in endothelial cells. Lab Investig 86(1):9–22

Dauphinee SM, Voelcker V, Tebaykina Z, Wong F, Karsan A (2011) Heterotrimeric Gi/Go proteins modulate endothelial TLR signaling independent of the MyD88-dependent pathway. Am J Physiol Heart Circ Physiol 301(6):H2246–H2253

De Chiara G, Marcocci ME, Torcia M, Lucibello M, Rosini P, Bonini P, Higashimoto Y, Damonte G, Armirotti A, Amodei S, Palamara AT, Russo T, Garaci E, Cozzolino F (2006) Bcl-2 phosphorylation by p38 MAPK: identification of target sites and biologic consequences. J Biol Chem 281(30):21353–21361

Desch CE, O'Hara P, Harlan JM (1989) Antilipopolysaccharide factor from horseshoe crab, Tachypleus tridentatus, inhibits lipopolysaccharide activation of cultured human endothelial cells. Infect Immun 57(5):1612–1614

Di Sano F, Ferraro E, Tufi R, Achsel T, Piacentini M, Cecconi F (2006) Endoplasmic reticulum stress induces apoptosis by an apoptosome-dependent but caspase 12-independent mechanism. J Biol Chem 281(5):2693–2700

Diette GB, Accinelli RA, Balmes JR, Buist AS, Checkley W, Garbe P, Hansel NN, Kapil V, Gordon S, Lagat DK, Yip F, Mortimer K, Perez-Padilla R, Roth C, Schwaninger JM, Punturieri A, Kiley J (2012) Obstructive lung disease and exposure to burning biomass fuel in the indoor environment. Glob Heart 7(3):265–270

Digaleh H, Kiaei M, Khodagholi F (2013) Nrf2 and Nrf1 signaling and ER stress crosstalk: implication for proteasomal degradation and autophagy. Cell Mol Life Sci 70(24):4681–4694

Downey GP, Dong Q, Kruger J, Dedhar S, Cherapanov V (1999) Regulation of neutrophil activation in acute lung injury. Chest 116(1 Suppl):46S–54S

Eckle T, Fullbier L, Wehrmann M, Khoury J, Mittelbronn M, Ibla J, Rosenberger P, Eltzschig HK (2007) Identification of ectonucleotidases CD39 and CD73 in innate protection during acute lung injury. J Immunol 178(12):8127–8137

Eltzschig HK, Faigle M, Knapp S, Karhausen J, Ibla J, Rosenberger P, Odegard KC, Laussen PC, Thompson LF, Colgan SP (2006) Endothelial catabolism of extracellular adenosine during hypoxia: the role of surface adenosine deaminase and CD26. Blood 108(5):1602–1610

Farley N, Pedraza-Alva G, Serrano-Gomez D, Nagaleekar V, Aronshtam A, Krahl T, Thornton T, Rincon M (2006) p38 mitogen-activated protein kinase mediates the Fas-induced mitochondrial death pathway in CD8+ T cells. Mol Cell Biol 26(6):2118–2129

Feoktistov I, Goldstein AE, Ryzhov S, Zeng D, Belardinelli L, Voyno-Yasenetskaya T, Biaggioni I (2002) Differential expression of adenosine receptors in human endothelial cells: role of A2B receptors in angiogenic factor regulation. Circ Res 90(5):531–538

Fleming IN, Elliott CM, Exton JH (1996) Differential translocation of Rho family GTPases by lysophosphatidic acid, endothelin-1, and platelet-derived growth factor. J Biol Chem 271:33067–33073

Ford ES, Murphy LB, Khavjou O, Giles WH, Holt JB, Croft JB (2015) Total and state-specific medical and absenteeism costs of COPD among adults aged ≥18 years in the United States for 2010 and projections through 2020. Chest 147(1):31–45

Fredholm BB, IJzerman AP, Jacobson KA, Klotz KN, Linden J (2001) International Union of Pharmacology. XXV. Nomenclature and classification of adenosine receptors. Pharmacol Rev 53(4):527–552

Fujita M, Kuwano K, Kunitake R, Hagimoto N, Miyazaki H, Kaneko Y, Kawasaki M, Maeyama T, Hara N (1998) Endothelial cell apoptosis in lipopolysaccharide-induced lung injury in mice. Int Arch Allergy Immunol 117(3):202–208

Gaipl US, Kuhn A, Sheriff A, Munoz LE, Franz S, Voll RE, Kalden JR, Herrmann M (2006) Clearance of apoptotic cells in human SLE. Curr Dir Autoimmun 9:173–187

Gill SE, Taneja R, Rohan M, Wang L, Mehta S (2014) Pulmonary microvascular albumin leak is associated with endothelial cell death in murine sepsis-induced lung injury in vivo. PLoS One 9 (2):e88501

Gill SE, Rohan M, Mehta S (2015) Role of pulmonary microvascular endothelial cell apoptosis in murine sepsis-induced lung injury in vivo. Respir Res 16:109

Giordano RJ, Lahdenranta J, Zhen L, Chukwueke U, Petrache I, Langley RR, Fidler IJ, Pasqualini R, Tuder RM, Arap W (2008) Targeted induction of lung endothelial cell apoptosis causes emphysema-like changes in the mouse. J Biol Chem 283(43):29447–29460

Goldar S, Khaniani MS, Derakhshan SM, Baradaran B (2015) Molecular mechanisms of apoptosis and roles in cancer development and treatment. Asian Pac J Cancer Prev 16(6):2129–2144

Groenendyk J, Michalak M (2005) Endoplasmic reticulum quality control and apoptosis. Acta Biochim Pol 52:381–395

Guan JL, Shalloway D (1992) Regulation of focal adhesion-associated protein tyrosine kinase by both cellular adhesion and oncogenic transformation. Nature 358(6388):690–692

Guan JL, Trevithick JE, Hynes RO (1991) Fibronectin/integrin interaction induces tyrosine phosphorylation of a 120-kDa protein. Cell Regul 2(11):951–964

Gustafsson AB, Gottlieb RA (2008) Recycle or die: the role of autophagy in cardioprotection. J Mol Cell Cardiol 44:654–661

Harrington EO, Lu Q, Rounds S (2007) Endothelial cell apoptosis. In: Endothelial biomedicine. Cambridge Press, Cambridge

Hengstermann A, Müller T (2008) Endoplasmic reticulum stress induced by aqueous extracts of cigarette smoke in 3T3 cells activates the unfolded-protein-response-dependent PERK pathway of cell survival. Free Radic Biol Med 44:1097–1107

Henriquez M, Armisén R, Stutzin A, Quest AF (2008) Cell death by necrosis, a regulated way to go. Curr Mol Med 8(3):187–206

Henson PM, Tuder RM (2008) Apoptosis in the lung: induction, clearance and detection. Am J Physiol Lung Cell Mol Physiol 294:L601–L611

Holguin F, Moss I, Brown LA, Guidot DM (1998) Chronic ethanol ingestion impairs alveolar type II cell glutathione homeostasis and function and predisposes to endotoxin-mediated acute edematous lung injury in rats. J Clin Invest 101(4):761–768

Hotchkiss RS, Strasser A, McDunn JE, Swanson PE (2009) Cell death. N Engl J Med 361 (16):1570–1583

Hsieh SJ, Zhuo H, Benowitz NL, Thompson BT, Liu KD, Matthay MA, Calfee CS, L. National Heart, N. Blood Institute Acute Respiratory Distress Syndrome (2014) Prevalence and impact of active and passive cigarette smoking in acute respiratory distress syndrome. Crit Care Med 42(9):2058–2068
Hu X, Yee E, Harlan JM, Wong F, Karsan A (1998) Lipopolysaccharide induces the antiapoptotic molecules, A1 and A20, in microvascular endothelial cells. Blood 92(8):2759–2765
Huang D, Khoe M, Befekadu M, Chung S, Takata Y, Ilic D, Bryer-Ash M (2007) Focal adhesion kinase mediates cell survival via NF-kappaB and ERK signaling pathways. Am J Physiol Cell Physiol 292(4):C1339–C1352
Hungerford JE, Compton MT, Matter ML, Hoffstrom BG, Otey CA (1996) Inhibition of pp125FAK in cultured fibroblasts results in apoptosis. J Cell Biol 135(5):1383–1390
Hynes RO (1992) Integrins: versatility, modulation, and signaling in cell adhesion. Cell 69 (1):11–25
Iizuka T, Ishii Y, Itoh K, Kiwamoto T, Kimura T, Matsuno Y, Morishima Y, Yamamoto M, Sekizawa K (2005) Nrf2-deficient mice are highly susceptible to cigarette smoke-induced emphysema. Genes Cells 10(12):1113–1125
Ilic D, Furuta Y, Kanazawa S, Takeda N, Sobue K, Nakatsuji N, Nomura S, Fujimoto J, Okada M, Yamamoto T (1995) Reduced cell motility and enhanced focal adhesion contact formation in cells from FAK-deficient mice. Nature 377(6549):539–544
Ilic D, Almeida EA, Schlaepfer DD, Dazin P, Aizawa S, Damsky CH (1998) Extracellular matrix survival signals transduced by focal adhesion kinase suppress p53-mediated apoptosis. J Cell Biol 143(2):547–560
Ilic D, Kovacic B, McDonagh S, Jin F, Baumbusch C, Gardner DG, Damsky CH (2003) Focal adhesion kinase is required for blood vessel morphogenesis. Circ Res 92(3):300–307
Imai K, Mercer BA, Schulman LL, Sonett JR, D'Armiento JM (2005) Correlation of lung surface area to apoptosis and proliferation in human emphysema. Eur Respir J 25(2):250–258
Iribarren C, Jacobs DR Jr, Sidney S, Gross MD, Eisner MD (2000) Cigarette smoking, alcohol consumption, and risk of ARDS: a 15-year cohort study in a managed care setting. Chest 117 (1):163–168
Jorgensen E, Stinson A, Shan L, Yang J, Gietl D, Albino AP (2008) Cigarette smoke induces endoplasmic reticulum stress and the unfolded protein response in normal and malignant human lung cells. BMC Cancer 8:229
Kasahara Y, Tuder RM, Taraseviciene-Stewart L, Le Cras TD, Abman S, Hirth PK, Waltenberger J, Voelkel NF (2000) Inhibition of VEGF receptors causes lung cell apoptosis and emphysema. J Clin Invest 106:1311–1319
Kasahara Y, Tuder RM, Cool CD, Lynch DA, Flores SC, Voelkel NF (2001) Endothelial cell death and decreased expression of vascular endothelial growth factor and vascular endothelial growth factor receptor 2 in emphysema. Am J Respir Crit Care Med 163:737–744
Kawasaki M, Kuwano K, Hagimoto N, Matsuba T, Kunitake R, Tanaka T, Maeyama T, Hara N (2000) Protection from lethal apoptosis in lipopolysaccharide-induced acute lung injury in mice by a caspase inhibitor. Am J Pathol 157(2):597–603
Kelsen SG, Duan X, Ji R, Perez O, Liu C, Merali S (2008) Cigarette smoke induces an unfolded protein response in the human lung: a proteomic approach. Am J Respir Cell Mol Biol 38:541–550
Kennedy SG, Wagner AJ, Conzen SD, Jordan J, Bellacosa A, Tsichlis PN, Hay N (1997) The PI 3-kinase/Akt signaling pathway delivers an anti-apoptotic signal. Genes Dev 11(6):701–713
Kerr J, Wyllie A, Currie A (1972) Apoptosis: a basic biological phenomenon with wide-ranging implications in tissue kinetics. Br J Cancer 26:239–257
Khaltaev N (2005) WHO strategy for prevention and control of chronic obstructive pulmonary disease. Exp Lung Res 31(Suppl 1):55–56
Khwaja A, Rodriguez-Viciana P, Wennstrom S, Warne PH, Downward J (1997) Matrix adhesion and Ras transformation both activate a phosphoinositide 3-OH kinase and protein kinase B/Akt cellular survival pathway. EMBO J 16(10):2783–2793

Kim SJ, Zhang Z, Hitomi E, Lee YC, Mukherjee AB (2006) Endoplasmic reticulum stress-induced caspase-4 activation mediates apoptosis and neurodegeneration in INCL. Hum Mol Genet 15 (11):1826–1834

Kim HP, Wang X, Chen ZH, Lee SJ, Huang MH, Wang Y, Ryter SW, Choi AM (2008) Autophagic proteins regulate cigarette smoke-induced apoptosis: protective role of heme oxygenase-1. Autophagy 4(7):887–895

Klionsky DJ et al (2016) Guidelines for the use and interpretation of assays for monitoring autophagy. Autophagy 12(1):1–222

Kramer K, Harrington EO, Lu Q, Bellas R, Newton J, Sheahan KL, Rounds S (2003) Isoprenylcysteine carboxyl methyltransferase activity modulates endothelial cell apoptosis. Mol Biol Cell 14(3):848–857

Kranenburg O, Poland M, Gebbink M, Oomen L, Moolenaar WH (1997) Dissociation of LPA-induced cytoskeletal contraction from stress fiber formation by differential localization of RhoA. J Cell Sci 110:2417–2427

LaCasse EC, Baird S, Korneluk RG, MacKenzie AE (1998) The inhibitors of apoptosis (IAPs) and their emerging role in cancer. Oncogene 17(25):3247–3259

Li H, Zhu H, Xu C, Yuan J (1998) Cleavage of BID by caspase 8 mediates the mitochondrial damage in the Fas-pathway of apoptosis. Cell 94:491–501

Lin JH, Li H, Yasumura D, Cohen HR, Zhang C, Panning B, Shokat KM, Lavail MM, Walter P (2007) IRE1 signaling affects cell fate during the unfolded protein response. Science 318 (5852):944–949

Linkermann A, Green DR (2014) Necroptosis. N Engl J Med 370(5):455–465

Lu Q (2008) Transforming growth factor-beta1 protects against pulmonary artery endothelial cell apoptosis via ALK5. Am J Physiol Lung Cell Mol Physiol 295(1):L123–L133

Lu Q, Rounds S (2009) Pulmonary endothelial cell death: implications for lung disease pathogenesis. In: Voelkel NF, Rounds S (eds) The pulmonary endothelium: function in health and disease. Wiley, Chichester, pp 243–260

Lu Q, Sakhatskyy P, Grinnell K, Newton J, Ortiz M, Wang Y, Sanchez-Esteban J, Harrington EO, Rounds S (2011) Cigarette smoke causes lung vascular barrier dysfunction via oxidative stress-mediated inhibition of RhoA and focal adhesion kinase. Am J Physiol Lung Cell Mol Physiol 301 (6):L847–L857

Lu Q, Newton J, Hsiao V, Shamirian P, Blackburn MR, Pedroza M (2012) Sustained adenosine exposure causes lung endothelial barrier dysfunction via nucleoside transporter-mediated signaling. Am J Respir Cell Mol Biol 47(5):604–613

Lu Q, Sakhatskyy P, Newton J, Shamirian P, Hsiao V, Curren S, Gabino Miranda GA, Pedroza M, Blackburn MR, Rounds S (2013) Sustained adenosine exposure causes lung endothelial apoptosis: a possible contributor to cigarette smoke-induced endothelial apoptosis and lung injury. Am J Physiol Lung Cell Mol Physiol 304(5):L361–L370

Machino T, Hashimoto S, Maruoka S, Gon Y, Hayashi S, Mizumura K, Nishitoh H, Ichijo H, Horie T (2003) Apoptosis signal-regulating kinase 1-mediated signaling pathway regulates hydrogen peroxide-induced apoptosis in human pulmonary vascular endothelial cells. Crit Care Med 31(12):2776–2781

Martin TR, Pistorese BP, Chi EY, Goodman RB, Matthay MA (1989) Effects of leukotriene B4 in the human lung. Recruitment of neutrophils into the alveolar spaces without a change in protein permeability. J Clin Invest 84(5):1609–1619

Martin TR, Hagimoto N, Nakamura M, Matute-Bello G (2005) Apoptosis and epithelial injury in the lungs. Proc Am Thorac Soc 2(3):214–220

Matot I, Weiniger CF, Zeira E, Galun E, Joshi BV, Jacobson KA (2006) A3 adenosine receptors and mitogen-activated protein kinases in lung injury following in vivo reperfusion. Crit Care 10(2):R65

Matsuzawa A, Ichijo H (2008) Redox control of cell fate by MAP kinase: physiological roles of ASK1-MAP kinase pathway in stress signaling. Biochim Biophys Acta 1780(11):1325–1336

Matthay MA, Ware LB, Zimmerman GA (2012) The acute respiratory distress syndrome. J Clin Invest 122(8):2731–2740

Matute-Bello G, Liles WC, Steinberg KP, Kiener PA, Mongovin S, Chi EY, Jonas M, Martin TR (1999) Soluble Fas ligand induces epithelial cell apoptosis in humans with acute lung injury (ARDS). J Immunol 163(4):2217–2225

Michaelson D, Silletti J, Murphy G, Peter DE, Rush M, Philips MR (2001) Differential localization of Rho GTPases in live cells: regulation by hypervariable regions and RhoGDI binding. J Cell Biol 152:111–126

Mizumura K, Cloonan SM, Nakahira K, Bhashyam AR, Cervo M, Kitada T, Glass K, Owen CA, Mahmood A, Washko GR, Hashimoto S, Ryter SW, Choi AM (2014) Mitophagy-dependent necroptosis contributes to the pathogenesis of COPD. J Clin Invest 124(9):3987–4003

Mizuno S, Yasuo M, Bogaard HJ, Kraskauskas D, Alhussaini A, Gomez-Arroyo J, Farkas D, Farkas L, Voelkel NF (2012) Copper deficiency induced emphysema is associated with focal adhesion kinase inactivation. PLoS One 7(1):e30678

Murray CJ, Atkinson C, Bhalla K, Birbeck G, Burstein R, Chou D, Dellavalle R, Danaei G, Ezzati M, Fahimi A, Flaxman D, Foreman S, Gabriel E, Gakidou N, Kassebaum S, Khatibzadeh S, Lim S, Lipshultz E, London S, Lopez AD, MacIntyre MF, Mokdad AH, Moran A, Moran AE, Mozaffarian D, Murphy T, Naghavi M, Pope C, Roberts T, Salomon J, Schwebel DC, Shahraz S, Sleet DA, Murray CJ, Abraham J, Ali MK, Atkinson C, Bartels DH, Bhalla K, Birbeck G, Burstein R, Chen H, Criqui MH, Dahodwala N, Jarlais DC, Ding EL, Dorsey ER, Ebel BE, Ezzati M, Fahami SF, Flaxman AD, Gonzalez-Medina D, Grant B, Hagan H, Hoffman H, Kassebaum N, Khatibzadeh S, Leasher JL, Lin J, Lipshultz SE, Lozano R, Lu Y, Mallinger L, McDermott MM, Micha R, Miller TR, Mokdad AA, Mokdad AH, Mozaffarian D, Naghavi M, Narayan KM, Omer SB, Pelizzari PM, Phillips D, Ranganathan D, Rivara FP, Roberts T, Sampson U, Sanman E, Sapkota A, Schwebel DC, Sharaz S, Shivakoti R, Singh GM, Singh D, Tavakkoli M, Towbin JA, Wilkinson JD, Zabetian A, Murray CJ, Abraham J, Ali MK, Alvardo M, Atkinson C, Baddour LM, Benjamin EJ, Bhalla K, Birbeck G, Bolliger I, Burstein R, Carnahan E, Chou D, Chugh SS, Cohen A, Colson KE, Cooper LT, Couser W, Criqui MH, Dabhadkar KC, Dellavalle RP, Jarlais DC, Dicker D, Dorsey ER, Duber H, Ebel BE, Engell RE, Ezzati M, Felson DT, Finucane MM, Flaxman S, Flaxman AD, Fleming T, Foreman K, Forouzanfar MH, Freedman G, Freeman MK, Gakidou E, Gillum RF, Gonzalez-Medina D, Gosselin R, Gutierrez HR, Hagan H, Havmoeller R, Hoffman H, Jacobsen KH, James SL, Jasrasaria R, Jayarman S, Johns N, Kassebaum N, Khatibzadeh S, Lan Q, Leasher JL, Lim S, Lipshultz SE, London S, Lopez AD, Lozano R, Lu Y, Mallinger L, Meltzer M, Mensah GA, Michaud C, Miller TR, Mock C, Moffitt TE, Mokdad AA, Mokdad AH, Moran A, Naghavi M, Narayan KM, Nelson RG, Olives C, Omer SB, Ortblad K, Ostro B, Pelizzari PM, Phillips D, Raju M, Razavi H, Ritz B, Roberts T, Sacco RL, Salomon J, Sampson U, Schwebel DC, Shahraz S, Shibuya K, Silberberg D, Singh JA, Steenland K, Taylor JA, Thurston GD, Vavilala MS, Vos T, Wagner GR, Weinstock MA, Weisskopf MG, Wulf S, Murray CJ, U. S. B. o. D. Collaborators (2013) The state of US health, 1990–2010: burden of diseases, injuries, and risk factors. JAMA 310 (6):591–608

No Authors (1997) Alpha 1-antitrypsin deficiency: memorandum from a WHO meeting. Bull World Health Organ 75(5):397–415

Noguchi T, Ishii K, Fukutomi H, Naguro I, Matsuzawa A, Takeda K, Ichijo H (2008) Requirement of reactive oxygen species-dependent activation of ASK1-p38 MAPK pathway for extracellular ATP-induced apoptosis in macrophage. J Biol Chem 283(12):7657–7665

Obeng EA, Boise LH (2005) Caspase-12 and caspase-4 are not required for caspase-dependent endoplasmic reticulum stress-induced apoptosis. J Biol Chem 280(33):29578–29587

Ola MS, Nawaz M, Ahsan H (2011) Role of Bcl-2 family proteins and caspases in the regulation of apoptosis. Mol Cell Biochem 351(1–2):41–58

Olson M, Kornbluth S (2001) Mitochondria in apoptosis and human disease. Curr Mol Med 1 (1):91–122

Parsons JT (2003) Focal adhesion kinase: the first ten years. J Cell Sci 116(Pt 8):1409–1416
Pastor LM, Sánchez-Gascón F, Girona JC, Bernal-Mañas CM, Morales E, Beltrán-Frutos E, Canteras M (2006) Morphogenesis of rat experimental pulmonary emphysema induced by intratracheally administered papain: changes in elastic fibres. Histol Histopathol 21(12):1309–1319
Perl M, Chung CS, Lomas-Neira J, Rachel TM, Biffl WL, Cioffi WG, Ayala A (2005) Silencing of Fas, but not caspase-8, in lung epithelial cells ameliorates pulmonary apoptosis, inflammation, and neutrophil influx after hemorrhagic shock and sepsis. Am J Pathol 167(6):1545–1559
Perl M, Chung CS, Perl U, Lomas-Neira J, de Paepe M, Cioffi WG, Ayala A (2007) Fas-induced pulmonary apoptosis and inflammation during indirect acute lung injury. Am J Respir Crit Care Med 176(6):591–601
Perl M, Lomas-Neira J, Venet F, Chung CS, Ayala A (2011) Pathogenesis of indirect (secondary) acute lung injury. Expert Rev Respir Med 5(1):115–126
Petrache I, Natarajan V, Zhen L, Medler TR, Richter AT, Cho C, Hubbard WC, Berdyshev EV, Tuder RM (2005) Ceramide upregulation causes pulmonary cell apoptosis and emphysema-like disease in mice. Nat Med 11:491–498
Petrache I, Fijalkowska I, Medler TR, Skirball J, Cruz P, Zhen L, Petrache HI, Flotte TR, Tuder RM (2006) Alpha-1 antitrypsin inhibits caspase-3 activity, preventing lung endothelial cell apoptosis. Am J Pathol 169:1155–1166
Pittet JF, Mackersie RC, Martin TR, Matthay MA (1997) Biological markers of acute lung injury: prognostic and pathogenetic significance. Am J Respir Crit Care Med 155(4):1187–1205
Qing DY, Conegliano D, Shashaty MG, Seo J, Reilly JP, Worthen GS, Huh D, Meyer NJ, Mangalmurti NS (2014) Red blood cells induce necroptosis of lung endothelial cells and increase susceptibility to lung inflammation. Am J Respir Crit Care Med 190(11):1243–1254
Reytor E, Perez-Miguelsanz J, Alvarez L, Perez-Sala D, Pajares MA (2009) Conformational signals in the C-terminal domain of methionine adenosyltransferase I/III determine its nucleocytoplasmic distribution. FASEB J 23(10):3347–3360
Rivo J, Zeira E, Galun E, Matot I (2004) Activation of A3 adenosine receptor provides lung protection against ischemia-reperfusion injury associated with reduction in apoptosis. Am J Transplant 4(12):1941–1948
Rounds S, Yee WL, Dawicki DD, Harrington E, Parks N, Cutaia MV (1998) Mechanism of extracellular ATP- and adenosine-induced apoptosis of cultured pulmonary artery endothelial cells. Am J Phys 275(2 Pt 1):L379–L388
Rubenfeld GD, Caldwell E, Peabody E, Weaver J, Martin DP, Neff M, Stern EJ, Hudson LD (2005) Incidence and outcomes of acute lung injury. N Engl J Med 353(16):1685–1693
Ruchko M, Gorodnya O, LeDoux SP, Alexeyev MF, Al-Mehdi AB, Gillespie MN (2005) Mitochondrial DNA damage triggers mitochondrial dysfunction and apoptosis in oxidant-challenged lung endothelial cells. Am J Physiol Lung Cell Mol Physiol 288(3):L530–L535
Rutkowski DT, Arnold SM, Miller CN, Wu J, Li J, Gunnison KM, Mori K, Sadighi Akha AA, Raden D, Kaufman RJ (2006) Adaptation to ER stress is mediated by differential stabilities of pro-survival and pro-apoptotic mRNAs and proteins. PLoS Biol 4(11):e374
Sakhatskyy P, Gabino Miranda GA, Newton J, Lee CG, Choudhary G, Vang A, Rounds S, Lu Q (2014) Cigarette smoke-induced lung endothelial apoptosis and emphysema are associated with impairment of FAK and eIF2alpha. Microvasc Res 94:80–89
Sandford AJ, Weir TD, Pare PD (1997) Genetic risk factors for chronic obstructive pulmonary disease. Eur Respir J 10(6):1380–1391
Sata M, Takabatake N, Inoue S, Shibata Y, Abe S, Machiya J, Wada T, Ji G, Kido T, Matsuura T, Muramatsu MA, Kubota I (2007) Intronic single-nucleotide polymorphisms in Bcl-2 are associated with chronic obstructive pulmonary disease severity. Respirology 12:34–41
Savill J, Dransfield I, Gregory C, Haslett C (2002) A blast from the past: clearance of apoptotic cells regulates immune responses. Nat Rev Immunol 2(12):965–975
Schaller MD, Hildebrand JD, Shannon JD, Fox JW, Vines RR, Parsons JT (1994) Autophosphorylation of the focal adhesion kinase, pp125FAK, directs SH2-dependent binding of pp60src. Mol Cell Biol 14(3):1680–1688

Schroder M, Kaufman RJ (2005) The mammalian unfolded protein response. Annu Rev Biochem 74:739–789

Shapiro SD (1995) The pathogenesis of emphysema: the elastase:antielastase hypothesis 30 years later. Proc Assoc Am Physicians 107(3):346–352

Shapiro SD (1999) The macrophage in chronic obstructive pulmonary disease. Am J Respir Crit Care Med 160(5 Pt 2):S29–S32

Shapiro SD, Goldstein NM, Houghton AM, Kobayashi DK, Kelley D, Belaaouaj A (2003) Neutrophil elastase contributes to cigarette smoke-induced emphysema in mice. Am J Pathol 163(6):2329–2335

Shen TL, Park AY, Alcaraz A, Peng X, Jang I, Koni P, Flavell RA, Gu H, Guan JL (2005) Conditional knockout of focal adhesion kinase in endothelial cells reveals its role in angiogenesis and vascular development in late embryogenesis. J Cell Biol 169(6):941–952

Simonis G, Wiedemann S, Joachim D, Weinbrenner C, Marquetant R, Strasser RH (2009) Stimulation of adenosine A2b receptors blocks apoptosis in the non-infarcted myocardium even when administered after the onset of infarction. Mol Cell Biochem 328(1-2):119–126

Sipkens JA, Hahn NE, Blom HJ, Lougheed SM, Stehouwer CD, Rauwerda JA, Krijnen PA, van Hinsbergh VW, Niessen HW (2012) S-Adenosylhomocysteine induces apoptosis and phosphatidylserine exposure in endothelial cells independent of homocysteine. Atherosclerosis 221(1):48–54

Steinberg KP, Milberg JA, Martin TR, Maunder RJ, Cockrill BA, Hudson LD (1994) Evolution of bronchoalveolar cell populations in the adult respiratory distress syndrome. Am J Respir Crit Care Med 150(1):113–122

Subramanian S, Steer CJ (2010) MicroRNAs as gatekeepers of apoptosis. J Cell Physiol 223 (2):289–298

Surolia R, Karki S, Kim H, Yu Z, Kulkarni T, Mirov SB, Carter AB, Rowe SM, Matalon S, Thannickal VJ, Agarwal A, Antony VB (2015) Heme oxygenase-1-mediated autophagy protects against pulmonary endothelial cell death and development of emphysema in cadmium-treated mice. Am J Physiol Lung Cell Mol Physiol 309(3):L280–L292

Szegezdi E, Fitzgerald U, Samali A (2003) Caspase-12 and ER-stress-mediated apoptosis: the story so far. Ann N Y Acad Sci 1010:186–194

Szegezdi E, Logue SE, Gorman AM, Samali A (2006) Mediators of endoplasmic reticulum stress-induced apoptosis. EMBO Rep 7(9):880–885

Tang K, Rossiter HB, Wagner PD, Breen EC (2004) Lung-targeted VEGF inactivation leads to an emphysema phenotype in mice. J Appl Physiol 97:1559–1566

Taraseviciene-Stewart L, Voelkel NF (2008) Molecular pathogenesis of emphysema. J Clin Invest 118(2):394–402

Thompson LF, Eltzschig HK, Ibla JC, Van De Wiele CJ, Resta R, Morote-Garcia JC, Colgan SP (2004) Crucial role for ecto-5′-nucleotidase (CD73) in vascular leakage during hypoxia. J Exp Med 200(11):1395–1405

Thorburn A (2004) Death receptor-induced cell killing. Cell Signal 16(2):139–144

Tuder RM, Wood K, Taraseviciene L, Flores SC, Voekel NF (2000) Cigarette smoke extract decreases the expression of vascular endothelial growth factor by cultured cells and triggers apoptosis of pulmonary endothelial cells. Chest 117(5 Suppl 1):241S–242S

Umapathy NS, Fan Z, Zemskov EA, Alieva IB, Black SM, Verin AD (2010) Molecular mechanisms involved in adenosine-induced endothelial cell barrier enhancement. Vasc Pharmacol 52 (5–6):199–206

Villar J, Kacmarek RM, Perez-Mendez L, Aguirre-Jaime A (2006) A high positive end-expiratory pressure, low tidal volume ventilatory strategy improves outcome in persistent acute respiratory distress syndrome: a randomized, controlled trial. Crit Care Med 34(5):1311–1318

Voelkel NF, Vandivier RW, Tuder RM (2006) Vascular endothelial growth factor in the lung. Am J Physiol Lung Cell Mol Physiol 290:L209–L221

Volmer JB, Thompson LF, Blackburn MR (2006) Ecto-5′-nucleotidase (CD73)-mediated adenosine production is tissue protective in a model of bleomycin-induced lung injury. J Immunol 176 (7):4449–4458

Wang C, Deng L, Hong M, Akkaraju GR, Inoue J, Chen ZJ (2001) TAK1 is a ubiquitin-dependent kinase of MKK and IKK. Nature 412(6844):346–351

Wang HL, Akinci IO, Baker CM, Urich D, Bellmeyer A, Jain M, Chandel NS, Mutlu GM, Budinger GR (2007) The intrinsic apoptotic pathway is required for lipopolysaccharide-induced lung endothelial cell death. J Immunol 179(3):1834–1841

Wang X, Cui L, Joseph J, Jiang B, Pimental D, Handy DE, Liao R, Loscalzo J (2011) Homocysteine induces cardiomyocyte dysfunction and apoptosis through p38 MAPK-mediated increase in oxidant stress. J Mol Cell Cardiol 52(3):753–760

Wheaton AG, Cunningham TJ, Ford ES, Croft JB, C. Centers for Disease and Prevention (2015) Employment and activity limitations among adults with chronic obstructive pulmonary disease–United States, 2013. MMWR Morb Mortal Wkly Rep 64(11):289–295

Wyatt AW, Steinert JR, Wheeler-Jones CP, Morgan AJ, Sugden D, Pearson JD, Sobrevia L, Mann GE (2002) Early activation of the p42/p44MAPK pathway mediates adenosine-induced nitric oxide production in human endothelial cells: a novel calcium-insensitive mechanism. FASEB J 16(12):1584–1594

Yakes FM, Van Houten B (1997) Mitochondrial DNA damage is more extensive and persists longer than nuclear DNA damage in human cells following oxidative stress. Proc Natl Acad Sci U S A 94(2):514–519

Zen K, Karsan A, Stempien-Otero A, Yee E, Tupper J, Li X, Eunson T, Kay MA, Wilson CB, Winn RK, Harlan JM (1999) NF-kappaB activation is required for human endothelial survival during exposure to tumor necrosis factor-alpha but not to interleukin-1beta or lipopolysaccharide. J Biol Chem 274(40):28808–28815

Zhande R, Dauphinee SM, Thomas JA, Yamamoto M, Akira S, Karsan A (2007) FADD negatively regulates lipopolysaccharide signaling by impairing interleukin-1 receptor-associated kinase 1-MyD88 interaction. Mol Cell Biol 27(21):7394–7404

Zhou Y, Mohsenin A, Morschl E, Young HW, Molina JG, Ma W, Sun CX, Martinez-Valdez H, Blackburn MR (2009) Enhanced airway inflammation and remodeling in adenosine deaminase-deficient mice lacking the A2B adenosine receptor. J Immunol 182(12):8037–8046

Zhou Y, Murthy JN, Zeng D, Belardinelli L, Blackburn MR (2010) Alterations in adenosine metabolism and signaling in patients with chronic obstructive pulmonary disease and idiopathic pulmonary fibrosis. PLoS One 5(2):e9224

FSC
www.fsc.org
MIX
Papier | Fördert gute Waldnutzung
FSC® C083411

Kolibri 360 GmbH
Ferdinand-Jühlke-Straße 7
99095 Erfurt, Deutschland
produktsicherheit@kolibri360.de